Land, Food and
Rural Development
in North Africa

Land, Food and Rural Development in North Africa

M. Riad El-Ghonemy

Routledge
Taylor & Francis Group

LONDON AND NEW YORK

First published 1993 by Westview Press, Inc.

Published 2018 by Routledge
52 Vanderbilt Avenue, New York, NY 10017
2 Park Square, Milton Park, Abingdon, Oxon OX14 4RN

Routledge is an imprint of the Taylor & Francis Group, an informa business

Copyright © 1993 Taylor & Francis

Library of Congress Cataloging-in-Publication Data
El-Ghonemy, Mohammad Riad, 1924–
    Land, food and rural development in North Africa / by M. Riad
El-Ghonemy.
        p.   cm. — (WVSS in social, political, and economic
development)
    Includes bibliographical references and index.
    ISBN 0-8133-8556-3
    1. Rural development—Africa, North.   2. Rural development—
Africa, North—Religious aspects—Islam.   3. Land tenure—Africa,
North.   4. Food supply—Africa, North.   I. Title   II. Series:
Westview special studies in social, political, and economic
development.
HN781.Z9C64   1993
307.1'412'0961—dc20   92-38895
                                                                           CIP

British Library Cataloguing in Publication Data
A CIP catalogue record for this book is available from the British Library.

ISBN 13: 978-0-367-00799-7 (hbk)
ISBN 13: 978-0-367-15786-9 (pbk)

To the memory of my father

Riad Omar

who taught me to learn from the fellaheen

Contents

Tables and Figures

Figures

Preface

Writing this book resulted from nearly 30 years of country studies that began in the early 1960s, when I was working with the UN Food and Agriculture Organization (FAO). From 1962 to 1970, I was closely associated -- in an advisory capacity -- with Libya's initial rural development program, based upon the resettlement of ex-Italian farms. This work led to holding a seminar on land policy, in September 1965, which brought together senior officials and academics from North Africa. Later in the 1970s, from the FAO Near East Office in Cairo, I had the opportunity to understand and assess the countries' experience in rural development, particularly their agrarian reform programs. The work was discussed in a regional meeting of Ministers of Agriculture held in Tunis in 1976. (The list of studies appears in Appendix 5.)

This book springs from four concerns. The first arises from the inadequate attention given to North Africa in development literature. The tendency has been to focus on Sub-Saharan Africa and to include North Africa in the broad geographical amalgams of the large Middle East region and the Mediterranean zone. In this way, the peculiarities of North African developmental problems, rural systems, and each country's conditions of scarcities, poverty and food insecurity tend to be obscured.

The second is the dilemma in rural-agricultural development generated by the structural adjustment programs adopted. Such hurried programs brought into the open the greatly imbalanced resource allocation, rewards and opportunities between the stagnant rural and booming urban sectors. Discussion of the identified dilemmas is of particular relevance in the context of the present debate on the roles of the State and the market in resource use and the distribution of land property and income, and, significantly, in alleviating rural poverty. It is also important because the economic reforms presently underway will shape the countries' rural-agricultural development policies for the years ahead.

The third concern is about falling food productivity, leading to a rapidly increasing reliance on food imports and the politically vulnerable food aid. The fast, cancerous growth of population (among the fastest in the world) coupled with the neglect of both irrigation investment and the food-producing sector, particularly in rain-fed areas, has increased the countries' degree of food insecurity and the numbers of the rural poor.

Lastly, I am worried about the consequences for social stability of the vigorous revival of Islamic extremism (widely known in Western media as fundamentalism) and governments' repressive responses at a time of their introduction of economic liberalization policy. Candidly, the movement is gaining wide support among the rural poor, the middle-class intelligentsia and some members of the armed forces. The activists' call for reforming the social order according to Islamic principles brought a repressive response from governments as well as fear and suspicion from Western countries. The latter consider this lusty movement to be a mobilizing force that may lead to a theocracy with ideological conflict resembling the Iranian experience. Irrespective of different interest-serving motives and forms of Islamic extremism, this movement is viewed in this study as an internally generated articulation of discontent with the moral substance of state policy regarding the application of Islamic principles to the remedy of such evils as gross inequalities, hunger, poverty, exploitative transactions, wasteful public consumption and corruption in bureaucracy. Governments tend not to trouble themselves with these evils, for which they are to be held accountable. They also seem to be unwilling to appreciate the legitimate call for pluralism in policy-making, based upon ballots and not bullets.

The four broad areas of concern are narrowed into concrete policy issues outlined in Chapter 1. Their linkages with rural development and food insecurity are examined in Chapters 7 and 8.

The appearance of this book resulted from several sources of help and encouragement. At government departments visited and rural areas studied, officials and farmers provided most of needed information. Within Queen Elizabeth House, Oxford University, the late Tony Mollett read the drafts of the first two chapters and offered helpful comments. His sudden death deprived me from receiving his views on the final version. I am indebted to Godfrey Tyler for reading, with patient care, Chapters 1, 2, 3 and 7 and for his useful criticisms and suggestions for improvement. The librarian Sheila Allcock and her able team, Bob Townsend, Gill Short, Liu Xin and Roger Grosvenor, were extremely helpful in the search for numerous reference materials. Likewise, Diane Ring, the librarian of the Middle East Centre, St. Antony's College, also of Oxford University, was very helpful, particularly in locating several references in Arabic.

In Cairo, I was saved from errors in my interpretation of Islamic arrangements for landownership inheritance by the wisdom of the late Shaikh Abdel-Monem El-Nemr and the comments received from Shaikh Mohamad Mobārak of al-Azhar University, Saad Hamza of Cairo University, the expert in law Abdel-Mageed Dekhale and from my brother Ahmad. I am very grateful for their clarification. During my seminar at Cairo American University (AUC), the graduate students of "Economics of the Middle East" raised probing questions about my presentation of the findings on distributional aspects of this study. I benefited from their remarks and those of their lecturer Morād Wahba. The kind assistance of the AUC library staff is appreciated.

I owe special gratitude to my former colleagues at FAO who provided me with much material. My deep appreciation to James Riddell, Hans Meliczek, Clifton Morojele, Fahmi Bishai, Ellen Kern, David Norse and Colleen McGowan-Moroni. At the Regional Office for the Near East in Cairo, Samir Miladi generously helped with the material on food consumption and nutrition. The librarian Laila Abdel-Gawãd provided me with information on the countries' recent development plans and the results of their census of agriculture.

I am indebted to the publisher, Westview Press. I have greatly benefited from the anonymous reviewers who, among other things, pointed out the importance of discussing the Islamic moral values in policy formulation, in rural people's daily transactions, and in landed property inheritance. I am grateful to Senior Acquisitions Editor Kellie Masterson and to Ellen McCarthy for her great patience with my frequent delays.

I have drawn on my earlier writings, including an article published in *Journal of Agricultural Economics* (1992) and a study prepared for the United Nations Economic Commission for Africa (1989). I am grateful for permission to use materials from these writings, and to Unwin Hyman for permission to reprint the map on North Africa.

Needless to say, none of the above scholars, officials, institutions and organizations has any responsibility for the contents of this book. I am solely accountable and bear responsibility for any deficiencies which still remain.

My deep thanks to Jane Norgrove, Rick Gaul and Kevin Tomes for wordprocessing, typesetting and for the layout of the numerous tables in the several versions of the manuscript. Lastly, my wife Marianne has patiently read the barely legible drafts and corrected the linguistic errors. She has gracefully put up with my long absences and solitary work at night and during the weekends. She helped me most with her deep understanding of my preoccupation throughout the gestation of this book.

Riad El-Ghonemy

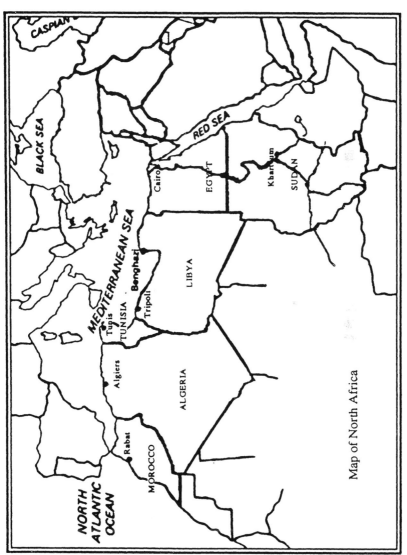

The map is reduced and adapted for printing. Source: Hopwood, Derek, *Egypt: Politics and Society, 1945-1990*, Harper Collins Academics, 1991. Reprinted by permission.

1

Introduction

This book explores the causal antecedents of current land tenure, food security and rural development problems in North Africa and the effects of approaches followed in tackling them. It has, therefore, a broad scope based on empirical findings and policy and helps to understand the lessons of the past.

In this inquiry, emphasis is placed on three important but controversial subjects. The first is the roles of the State and the market in the distribution of land property rights: enhancing food production, alleviating rural poverty, and in the expansion of aggregate supply of cultivable land through irrigation investment. The second is food security related to the hydraulic character of agriculture, rapid population growth and hunger alleviation *via* secured access to productive land and intensive family labor use. The third is the implication of the vigorous revival of the Islamic moral force for policy-making, with respect to the elimination of injustice in property and production relations[1], and the abolition of exploitative economic transactions. We also attempt to clarify the Islamic principles *(Shari'ah)*[2] concerning justice, exploitation, poverty and inheritance of landed property and their effects on rural development.

The Principal Issues

Irrigation, Investment and Food Security

In the semi-arid climate of the North African region, food security of the agricultural population, who are mostly cereal growers, depends chiefly on a combination of land tenure arrangements and the hydraulic characteristics of agriculture. The latter refers to the scarcity of water supply for irrigation, rainfall instability and to the extent of permanently irrigated lands, especially for staple food production.

Irrigation investment is, therefore, crucial to augmenting the aggregate supply of productive land and to reducing instability of both food production and the flow of farmers' income. Moreover, irrigation investment determines the rate of absorption of labor force in agriculture through increasing the intensity of land and labor use. But irrigation is not a politically neutral technological change. In practice, public investment in irrigation is caught in each country's balance of political power in general, and landownership-based lobbying strength in particular, determining who benefits.

The study views economic security of the *fellaheen*[3] and the pastoral nomads in terms of their ability to command their food intake over the whole year. We consider the acquisition of the *fellaheen's* nutritional requirements not by way of dependence on an imperfect grain- market, bureaucratic distribution of food and foreign donation. Rather, it is through self-produced food from legally-secured rights in landholding supplemented by earnings from non-farm activities. In this sense, economic security of the household becomes synonymous with a combination of secured family-labor employment and command over food intake. For the nomads who are the highest risk bearers, the minimum economic security is to be realized through livestock ownership, accessible employment opportunities in settled agriculture together with the legal protection of their legitimate grazing rights and access to water. These rights are to be guarded by the State against encroachment by land speculators.

For both groups, the study considers state provision of social security in terms of primary health service and education as a citizenship right, enhancing their productive abilities and enabling them to have a choice of employment opportunities. Thus, rural household-food security and the ability to choose between alternative opportunities are primary elements of rural development for escaping poverty.

Duality, Inequality and Social Instability

There are some bitter lessons to be learnt from past experience of North African countries. Available information on poor households' consumption in rural areas contrasts sharply with the affluent style of life seen in cities. There are indications of increasing inequalities between rural and urban quality of life. Within rural areas, a distinctive duality exists between the relatively small irrigated sector and the large rainfed sector. Governments' own policies for the allocation of resources among irrigated and rainfed areas, and their planned pattern of agricultural growth have invariably perpetuated this duality and widened the income gap within the agricultural sector. It is true that historical record indicates that this duality was initiated during the colonial administration. But, after nearly five decades of political independence, colonial rule is not to blame for the *present* conditions of rural poverty, gross inequality and food insecurity which threaten social stability.

Disaggregated results of recently conducted household expenditure surveys show a prevalence of undernutrition, absolute poverty and illiteracy among the cash wage-dependent landless workers, insecure tenants, small farmers in rainfed areas, artisanal fishermen and nomadic people. Most of these disadvantaged poor people are also the victims of documented consequences of prolonged droughts, leading to human disaster of starvation and high mortality during famine (e.g. Sudan in 1984-1985 and 1990- 1991).

If hunger and poverty of these masses were tolerated in the past, governments cannot, in the years ahead, afford the consequences of social instability. The threat of social (and political) instability is likely to increase if the present limitation of political freedom and the rights to organize free agricultural trade unions continues. In particular, it tends to be compounded by the lack of effective participation of the *fellaheen* in programing and implementing rural development schemes. Paternal control by state bureaucracy has invariably tended to ignore the *fellaheen's* perceptions, and inhibit their motivations and enterprising abilities. Moreover, being politically frustrated and strict observants of Islamic teachings, the rural poor tend to follow the movement of Islamic fundamentalists, questioning the legitimacy of many existing laws and institutions, and calling for the elimination of corruption in government bureaucracy and all forms of fraud and exploitation in transactions.

A Political Economy Perception

This book intends to examine the political economy of rural development in six North African Arab countries with an historical perspective. As its title suggests, it views state policy on the reallocation of landed property rights as the principal ingredient of anti-poverty rural development. It aims at the removal of institutional barriers and the realization of food security at both household level and the country as a whole. In this perception, we recognize that land policy does not in itself achieve these results. In a dynamic national economy, it is a complex process and politically based. Moreover, rural development strategies vary widely in aims, content, pace and scale of implementation, as do results. We are also aware of the fact that decisions on redistributive land policy and anti-poverty rural development are political, and they hinge on the structure of land-based power relations.

Structural Linkages. Another important facet of the complex rural development process of reducing inequality, hunger and poverty is that the realization of these benefits is not independent of the structural factors operating in the national economy. For a simplified illustration, these interdependent factors are summarized and schematically presented in Figure 1.1. Before explaining, in the next chapter, the reciprocal relationship between these factors and discussing the conceptual issues behind them as applied to North Africa, it is worth referring to them. In broad terms and given each country's natural resource endowment and ideology, they include: rates of investment and

FIGURE 1.1 Dynamics of Land Policy, Rural Development Process and Poverty

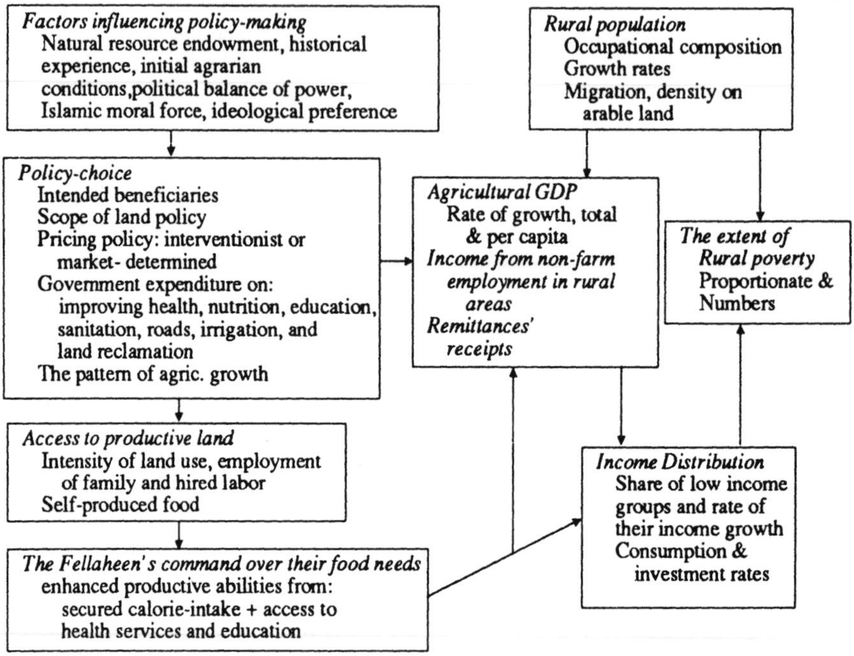

Factors influencing policy-making Natural resource endowment, historical experience, initial agrarian conditions,political balance of power, Islamic moral force, ideological preference	**Rural population** Occupational composition Growth rates Migration, density on arable land

Policy-choice
Intended beneficiaries
Scope of land policy
Pricing policy: interventionist or
    market- determined
Government expenditure on:
    improving health, nutrition, education,
    sanitation, roads, irrigation, and
    land reclamation
The pattern of agric. growth

Agricultural GDP
Rate of growth, total
& per capita
*Income from non-farm
employment in rural
areas
Remittances'
receipts*

The extent of
Rural poverty
Proportionate &
Numbers

Access to productive land
Intensity of land use, employment
of family and hired labor
Self-produced food

Income Distribution
Share of low income
groups and rate of
their income growth
Consumption &
investment rates

The Fellaheen's command over their food needs
enhanced productive abilities from:
secured calorie-intake + access to
health services and education

population growth as well as the dynamics of demand for both labor outside agriculture and agricultural products. They also include the choice between pricing by interventionist policy and by market mechanism, and between budgetary allocation for a balanced development of the agricultural and manufacturing sectors and for military expenditure. Besides, there are decisions to be made on the proportional allocations of public spending on roads, health and education among and within rural and urban areas. Within the agricultural sector, there are policy decisions to be made on the allocation of public investment among areas growing staple food and exportable crops, and between irrigated areas and the hitherto neglected rainfed traditional subsector. There are also decisions to be made by governments and private investors on the type of technological change (capital intensive or labor-using), depending upon the character of the rural economy.

There is nothing strikingly new about these linkages. They have been established theoretically and empirically.[4] There is a broad agreement that those who own, or have a secured access to, productive land and education, particularly primary schooling, benefit from agricultural growth and respond to technological change. Cash wage-dependent landless workers, on the other hand, having no alternative employment opportunities, tend to be net-food buyers, involuntarily unemployed part of the year and at high risk of undernutrition.

Agricultural Growth and Rural Poverty. What we want to explore from empirical evidence is the relationship between inequality, rural poverty and agricultural output growth in general, and food in particular. The presupposition is that by reliance on agricultural growth *alone* without reducing land concentration, it would require a very long time for most of the present generation of rural poor to rise above the poverty line so defined by each country. Otherwise, at a fixed distribution of land, the market-generated high rate of growth of agricultural domestic product per labor unit must be sustained sufficiently long to raise the average income of the poor above the poverty line as insinuated by the founder of economic thought, Adam Smith (1776). Smith said: "It is not the actual greatness of national wealth, but its continual increase, which occasions a rise in the wages of labor" (The Wealth of Nations, chapter 8). This paradigm has, since then, remained the conventional prescription of the anti-state intervention economists, particularly the followers of Ricardo.

It follows that if agricultural output growth is the prime aim of rural development where land distribution is highly skewed, landlords' profits from market-determined high rents rise and the share of tenants and hired workers in the expanded total output diminishes and poverty inevitably becomes normal. In other words, where economic activities in agriculture and the distribution of income are determined by an imperfect market mechanism, neither the landlords, irrigation pump-owners and traders, nor the State are to blame for the increasing inequalities and the fate of the rural poor. In such situations, the affluence of the former and the poverty of the masses of workers, small tenants and nomadic groups become in the nature of things.

It would be wrong to imagine that the countries of North Africa have not pursued policies to widen the opportunities for secure access to land and food in rural areas. On the contrary, in most of them there have been, since the 1950s, remarkable adjustments to old agrarian institutions. It would be equally incorrect to imagine that poor *fellaheen* and nomadic people are detached from the market and technological change. Despite existing climatic and institutional constraints, they are actively involved in the domestic and international labor market as well as in the exchange economy. They also adopt whatever technology is accessible and suitable to their systems of production.

Our concern over persistent inequalities, undernutrition and poverty in North Africa expressed earlier springs from the fragmentary achievements of rural development macro-policies which invariably fell short of the high expectations raised by the politicians promises at the time of their inception, particularly with regard to land reforms. Governments have also failed to achieve self-sufficiency in food production, and to reduce significantly dependence on food imports and aid to feed their own people. Nevertheless, compared to the pre- 1950s pervasive deprivation, a diverse proportion of the *fellaheen* have gained access to land, subsidized complementary inputs and to primary education and health care services. These gains are unlikely to be sustained in the 1990s, owing partly to increasing pressure on the scarce cultivable land, and partly to governments' financial difficulties in the 1980s, leading to their adoption of the World Bank and the IMF's induced economic reforms. With expected major cuts in subsidies and public expenditure on social services and land reclamation, combined with price support for export crops to the disadvantage of cereals, income distribution in rural areas and food security are likely to worsen.

The foregoing is a sample of the crucial issues to be explored. Also, the ambiguity of the frequently used key terms 'land policy', 'land tenure', 'agrarian system', 'balance of political power', 'exploitation', 'food security', and 'rural development' will be clarified in Chapter 2. Likewise, the several interpretations of relevant Islamic principles are briefly presented in the course of the study.

Questions Addressed

What has been said so far suggests a number of hypotheses on several linkages which need to be empirically verified. In particular, the following principal questions are addressed:

1. How can shifts in the balance of political power determine the content of land policy and the design of rural development strategy? In what way has the food-producing sector been affected by policy change?
2. What form of government would effectively introduce major pro-*fellaheen* changes in landed property rights and the distribution of income in rural areas?
3. Under what conditions does state-planned pattern of agricultural growth perpetuate inequalities and dependence on food imports?
4. What are the economic effects of government-administered land allocation among food and non-food crops and pricing of land values and agricultural products?
5. Is the choice of technological change influenced by the land-based lobbying structure? How have land tenure arrangements responded to irrigation expansion? What institutional arrangements have the farmers and nomads made for risk-aversion and for collective self-insurance against uncertainties of rainfall and market forces?
6. Is nutritional status associated with landholding? Is the food-grain production associated with the size of landholding?
7. Is the region's past famine institutionally-based, and has the institutional setting created a human disaster out of prolonged droughts?
8. Given the countries' diverse economic structure and agro-climatic characteristics, are the rates of agricultural growth and levels of landlessness and poverty associated with the degree of inequality in the size distribution of land? Would the break-up of large farms for

redistribution in small farms disrupt total agricultural and food production?

9. Are the Islamic principles of inheritance to blame for the continuing parcelation of land property from one generation to the next?

10. What are the dilemmas in rural development generated by structural adjustment programs?

The Approach

This book is written for the general reader, the policy-maker and analyst, as well as students of agricultural and rural development in Africa and the Middle East. The study enables the reader to understand the unique importance which the institutions of land tenure and State authority, inter-acting with Islamic culture have in influencing resource use, the distribution of wealth, income and land-based political power.[5] The analytical procedure alerts students of development to the consequential erroneous view occurring when the political, moral and institutional factors are ignored, or taken for granted on the assumption of institution- neutrality.[6] Thus, a partial understanding, leading often to a distorted policy recommendation results, if land tenure arrangements and land-based power structure are neglected or avoided by governments and donors, as 'sensitive issues'.

Understanding Institutions and History

For the adequate study of rural development problems, institutional arrangements in a specific agrarian system, Islamic moral values, and lobbying structure cannot be ignored, as unpredictable and unquantifiable variables. Nor can their impact be adequately appreciated if they are treated like weather, as external factors. They are necessary for understanding the social organization of the rural economy. In our study, moral and institutional factors are considered as integral factors influencing policy choice, agricultural production, rural employment, and the distribution of income and consumption.

By institutional arrangements in the agrarian system is meant the legal and customary rights of ownership and use of productive assets. They include: granting land to individuals or corporations by the State; inheritance of property and use rights; property transfer through inter-family marriage; grabbing land by virtue of social power and official status; and grazing rights in communally held land. They also comprise leasing arrangements, redistribution of landed-property rights and regulation of rental values enforced by law through government intervention. Likewise, they include custom-determined arrangements for both credit and labor use in agriculture.[7]

A full understanding of the institutional arrangements makes it possible to examine the *fellaheen's* response to economic factors such as technological change, risk-aversion and market uncertainty. By way of illustration, the problems of the credit market, the allocative efficiency in the steadily increasing small fragmented farming units and the preference of small producers for informal sources of credit are better perceived if the economist is aware of the Islamic principles of inheritance and the prohibition of *riba* (usury) in borrowing money. There are also the customs behind female labor participation and the large family size together with its sex composition.[8] Understanding the land-based lobbying structure and alliance of interests between private capitalists and bureaucrats, influencing the allocation of land and irrigation investment among landholding groups, and determining who benefits and by how much is another example.

In addition to the importance of institutions, history matters. It enables us to understand the causal antecedents of present land concentration and the duality of production structure into modern and traditional subsectors. History is also important in understanding how the content of rural development strategy, particularly its land policy component, mirrors the shifts in the balance of political power at different points in time. Moreover, the historical approach enables us to understand how the political and economic powers of the pastoral tribes have been weakened through the progressive alienation of their grazing lands and limitation of their mobility by the "modern" urban power as Ibn-Khaldoun, the fountainhead of the philosophy of human civilization history, forecast six centuries ago.

Within an historical perspective, the analysis of social organization of the rural economy combines a monographic treatment with a quantitative analysis of some production and distributional variables, employing simple statistical techniques. Nevertheless, the primary focus of the discussion is on state actions intended to correct the market failure, and on empirical events that explain them and their consequences. For this purpose, the study examines problems and policies at two levels: regional (chapters 3 and 7); and country case studies (chapters 4, 5 and 6).

Analytical Difficulties

Recognizably, our inquiry faces methodological and measurement problems. There are difficulties in the interpretation of the ideology behind the formulation of land policy and rural development strategy. Likewise, stated policy objectives are often ambiguous and loaded with emotive words. In North Africa, publicly declared aims are merely an expression of a mixture of nationalism and concern about rural underdevelopment problems. What is in the minds of the country leadership may be different from what is stated in development plans and presented by state organs in pursuasive political terms.

The above paragraph raises two questions which face the analyst with a dilemma. One is whether the objectives of rural development strategy and land policy -- being ambiguous -- be considered the starting point in the assessment?

Or, should we abstract them from what actually happened during the sequential stages of implementation? Though this derivation may have an objective connotation, it involves a value judgement influenced by our personal belief about what the objectives should be. The second best is to employ the derived objectives as complementary to the primary objectives explicitly stated by the country leadership at the time of policy declaration.

The other analytical question is how to measure the effects of land policy and related rural development activities in isolation from the dynamic process of national development in an international context? How can one reach the conclusion that the policy has or has not realized its objectives? How to be objective about measuring such vague descriptive connotations in policy aims as 'the *fellaheen's* dignity and liberation', 'abolition of feudalism and colonial power exploitation', 'social justice' . . .? As I understand the branches of social science, they do not possess objective criteria (free of ethical judgement) on which there is a substantial consensus about measuring these terms.[9]

We also encounter difficulties in relating statistical data by way of cross-sectional and time series analyses for the same country (e.g. land concentration, distribution of income or consumption, daily wage rates, rental values, nutritional status, and estimates of poverty level). One difficulty is that statistical data 'before' and 'after' a major policy are not always available, and, in some cases, they are not comparable across countries due to different definitions used. Moreover, they are usually at aggregate level, masking variations in localities and among land tenure groups.

Another difficulty is that considering the 'after' data as effects is in a strict sense misleading, because we do not precisely speak of the 'before' as causes. Perhaps what we can do is to use carefully our faculties and the professional tools at our disposal to identify and describe the prevailing food situation and agrarian relations without the introduced policies. We can, then, employ the policy elements as instruments of change analyzing the consequent equity, production and poverty conditions within an adequate time frame of at least 10-15 years after implementation. This approach brings another problem in analysis, i.e. the problem of overlapping events when we neatly break down historical experience into periods.

The Structure of the Study

The plan of the book is as follows. After this introductory chapter, we present the conceptual framework which clarifies the meaning of the key terms used in the study, the nature of their complementarities, and how they coalesce in one meaning: rural development. Chapter 3 introduces the reader to the regional setting and identifies the policy issues on the critical development problems. The chapter characterizes the salient common features, as well as the structural differences among the countries to the extent that they are directly related to rural development problems. The constraints affecting rural women's participation in agriculture are discussed.

The next three chapters comprize case studies. Each country study briefly traces the origin of the present agrarian system and rural development efforts instituted since the 1950s. The purpose is to understand how the characteristic features of rural underdevelopment shaped during colonial rule have influenced post-1950s policy choice, and pace of change. The effects of the recently adopted structural adjustment programs on food security, and the shares of agriculture, health and education in public spending and in GDP are briefly explored. The presentation of the six countries' experience follows their grouping into: the Maghreb in chapters 4 and 5 and the Nile Valley countries in chapter 6. The format of presentation is not uniform, and a few countries' experience are discussed in more detail according to availability of reliable data.

Through inter-country comparisons, Chapter 7 examines the distributional and rural welfare consequences of the countries' different paths of rural development strategies. It begins with an exploration of the main sources of agricultural growth, with emphasis on inter-country variations in technological change. The purpose is to assess the effects these policies have had on the pace of reducing inequalities in rural areas. In the course of exploring common problems of small fragmented farms, the controversial role of Islamic principles of inheritance in continuing parcelation of land is investigated. Subject to availability of comparable data, relationships between agricultural growth, nutrition, land concentration, poverty incidence and the quality of life indicators are analyzed.

The final chapter draws some conclusions having policy implications. After outlining the response of country planners and politicians to rural underdevelopment problems since the 1950s, the chapter discusses the dilemmas facing them in finding satisfactory solutions to short- and long-term rural and national development problems, particularly those generated by the economic reforms under way. The chapter also points to a set of broader questions that are closely related to the inquiry, and which are considered to have profound influence on policy-making in the years ahead.

Notes

1. By production relations, we mean the relations between and among land owners, tenants, share-croppers, hired agricultural workers, agro-pastoralists, irrigation pump-owners, state farm managers, creditors and debtors. This definition is an adaptation to that made by Rosenzweig, Binswanger and McIntire in Roland Lee, editor, (1988, p. 77).

2. *Shari'ah* is the whole fundamental Islamic teachings laid down in the Qur'an and explained by the Prophet Mohammad's sayings *(Hadith).* Moslem society should follow

these principles spiritually in relation with God and in relations between individuals described in the *Fiqh*, simply known in Western literature as Moslem law.

3. *Fellaheen* (pl. fellah), see 'Glossary'.

4. See for example Chenery *et al* (1974), Streeten *et al* (1981), Ghai and Radwan, eds (1983), Booth and Sundrum (1985), and El-Ghonemy (1990, Chapters 6 and 7).

5. On the importance of understanding these non-economic aspects, Joan Robinson of Cambridge University says, "Western teaching pretends to be scientific and objective by detaching the economic aspect of human life from its political and social setting; this distorts the problems that it has to discuss rather than illuminating them." See her penetrating study, *Aspects of Development and Underdevelopment*, p.3. Cambridge University Press, 1979.

6. On the responsibility of economists making these assumptions, Little says the economist "must remember that the given premises will be understood to run in terms of real individuals. But the economist himself works in terms of rather unreal abstractions; he must accordingly remember that his practical conclusions are suspect. The non-economist should be warned of the assumptions which the economist slips in, in order to be able to deduce his practical conclusions from the given premises" . . . (p. 127). . . . "Consequently, [the economist] must realize that he is always open to the suspicion that his political inclinations have colored his comment on the theory [of economic welfare]", p. 258 (1958, second edition, paperback), between brackets are added.

7. The importance of institutions and history in understanding economic problems has been stressed and argued by many development economists. In his *Asian Drama*, 1958, Gunnar Myrdal appealed for realism in theorizing economics and he formulated an alternative approach. For a comprehensive review of the main literature on this subject see Bardhan and Stiglitz in Bardhan (editor) 1989.

8. According to Islamic *Shari'ah*, land of a deceased owner is divided into shares among the heirs, whereby the male receives double the female's share. See a discussion of these inheritance principles in Chapter 7.

9. On this point of consensus in social science with regard to policy analysis, see for example: Little (1957) second edition, Chapter 5 "Value judgements and welfare Economics; and Johnston and Clark (1982), Chapter one "Policy Analysis and the Development Process".

2

The Conceptual Issues

This chapter sets the explanatory framework for understanding the subject of the study. It clarifies the meaning of some basic concepts behind policy choice and examines the linkages between them. The purpose is to comprehend the expected contribution of land policy and rural development efforts to the alleviation of food insecurity, and rural poverty through improved income distribution and sustained agricultural growth.

Land Tenure and Access to Land

Land tenure denotes institutional arrangements pertaining to property rights and obligations of the agricultural landholders. It indicates the division of production responsibilities and the distribution of shares in the produce of land and other means of production, among owners and users of land. These elements of contractual arrangements may be legally established, customarily practised for economic needs, or a combination of both. In agrarian societies like those covered by this study, land tenure arrangements are likely to be closely associated with the distribution of income and power, and social relations in rural areas.

For understanding the role of land tenure in rural development, three distinctions are made. The first is between the rights and responsibilities of who owns the land, and who actually operates it, and manages its use for agricultural purposes. Public or private *ownership* is, therefore, distinct from *holding*, particularly under landowner absenteeism. Holdings in the census of agriculture refer to owned or leased sizes of farms. While the size of holdings matter in resource use and its allocative efficiency, ownership matters in the understanding of land-based power in the society, the distribution of material wealth and income, and the right to transfer its property. Ownership also embodies the institutional arrangements of leasing-out, inheritance as well as land taxation. Since the dawn of Islam, *private* ownership of land has been associated with payment of land tax *(Ushr* and *Kharāj)*.

The second distinction to be made is between the physical meaning of landed property (its productive quality and soil texture), and its institutional content which links the productive land with the market. The latter embraces purchase, sale, lease, crop- sharing arrangements, mortgage, and the transaction costs involved. The productive quality and the institutional content of land rest chiefly on access to water supply which is scarce and crucial for food production in North Africa. Unlike land, water is mobile for irrigation and cost is incurred in moving it to land. Pasture lands using natural forage for grazing by the nomads are not liable to land taxation and have no scarcity value for transaction. In irrigated agriculture, access to land is, therefore, useless without secure access to water rights. Both are the key to access to credit and technological advance in production, and in turn, to the modernization of agriculture and the realization of land holders' food security. If labor use, marketing transactions, customary arrangements, property-based power relations and the legal framework are added to access to land, water, credit and technical knowledge, these interrelated elements can be termed *'agrarian system'*.

The third distinction to be made is between access to agricultural land through the market and non-market arrangements. Under the former, contractual transaction is arranged by bilateral bargaining between the seller or lessor (owner) and the buyer or lessee (tenant). Leasing land has been recognized by Islamic principles *(Ijara)* as a practical arrangement, serving economic needs. Within the market, we need to distinguish between the land lease-market and land-purchase (or sale) market. There are several submarkets, whereby land values (purchase price and rental value) are differentiated by location, density of rural population on land, size of farm, intensity of land use and cropping system. Non-market arrangements, on the other hand, refer to the transfer of land title or right to manage its use through inheritance, inter- family marriage, and land extortion by virtue of social power and official status. These arrangements include also granting land under concessional arrangements, negotiated transfer of landed property from ex-colonial to independent government, and state actions intended to remedy defective land tenure arrangements. This brings us to the meaning of land policy.

Ambiguities in Land Policy

There is an ambiguity in the use of the term 'land policy'. Academics and policy makers use the same term in different senses, having different content and empirical consequences. In what follows, we attempt to explain the difference in the practical expression of land policy.

In concrete terms, land policy does not mean only a state action for bringing new lands to cultivation through irrigation investment, soil conservation, construction of drainage network, and physical consolidation of fragmented parcels. These are technical changes. They would not bring about justice and distributive equity in situations where the distribution of land ownership is highly unequal, and the value of output continues to be inequitably shared. Technical change for augmenting the effective supply of land (actually cropped arable land by means of irrigation, reclamation and application of land-substitutes raising crop yield) is *complementary* to the *core* of land policy. The core is the amendment of iniquitous practices in land tenure. The amendment comprizes a change in the scope of management rights and accompanying incentives with or without a transfer of property rights. The purpose is the reduction of gross inequality of incomes and opportunities in having access to land. This, in turn, leads to the alleviation of poverty and undernutrition through the *fellaheen's* greater access to productive land and to self-produced food.

The Arabic language seems to be more precise in expressing this complementarity in one term *'Islāh Zirā'ee'*. It has a meaning distinct from settling farmers in newly reclaimed lands *(Istitān, Istislāh)*. In contrast, these alterations by public action have a double meaning in English language. The term 'agrarian reform' embraces a wide range of institutional and technical changes. Its meaning has been expanded, and is so ambiguous that it may omit the core of land policy. Usually it embraces one or more of the following: land settlement schemes in publicly-owned land, registration of land titles, rent control, lending credit to small farmers to purchase land in the open market, consolidation of fragmented holdings, cadastral surveys, land taxation and so forth. For political convenience, 'agrarian reform' policy may leave the existing concentration of landed property and the corresponding power structure unchanged. With land settlement schemes, as with tenancy regulation, social injustice and the incidence of rural poverty are not likely to be significantly reduced in the national context. The impact is more likely to be minimal where the size distribution of land, together with benefits from technological change and agricultural growth is very skewed. In such circumstances, there is no guarantee that the poor would benefit from general agricultural growth.

The nearest English term to the practical expression of the Arabic term *'Islāh Zirā'ee'* is 'land reform'. Both terms are fundamentally concerned with the realization of a greater equality of income distribution through a wider access to land and complementary inputs. Accordingly, land reform could be termed agrarian reform, but the converse need not be true. Land settlement schemes and consolidation of fragmented holdings form a part of land policy and are *not* an alternative to land reform. They are necessary public action in almost all countries of North Africa for extending the cultivated areas, rationalizing its use, providing additional employment opportunities and for increasing agricultural output.

However, both land settlement and consolidation schemes require heavy public expenditure for gradually bringing new lands (usually state-owned) into cultivation, for restructuring the size of holdings and the organizaton of production. They also require investment in social overhead capital (schools, clinics, roads, mosques or churches) which are public services needed for the rural communities formed. Furthermore, being part of the national budget and implemented by the bureaucratic system, land settlement and consolidation schemes are more vulnerable to budgetary cuts than the politically-committed redistributive land reform which is also not cost-free. Speedy implementation of land reform and the provision of the usually subsidized inputs incur expenditures from the national budget; hence the tax payers' concern about land reform.

The scale of the burden on the national budget depends upon the terms of payment for land by the benificiaries and compensation to be paid for expropriated property, particularly if it was owned by foreigners and paid in foreign exchange. Budgetary burden depends also on whether external resources are provided by foreign donors to enhance the implementation of land reform. Thus all types of land policy incur costs but for political and social pressure considerations, land reform tends to have an absolute priority for demanding financial support.

The Market and Redistribution of Land Property

Can the market mechanism through the private sector redistribute property rights in land for the benefit of the *fellaheen?* Recently, there has been a sharp swing away from the 1950s and 1960s international development thinking which assigned the State in developing countries an extensive role to amend the market failures in bringing about a greater equality in the distribution of income and opportunities.

Since the early 1980s, Western donor countries and international financing institutions revived the faith in the market. In their policy prescription to developing countries, they have emphasised (in some cases they even insisted) that the market mechanism freed from government intervention and planning is capable to redistribute property rights in land. Briefly, they argue with passion that this market approach ensures that resources in agriculture are used more efficiently by a strong private sector through enhanced investment. Those who want to own land can buy it at the market price from those willing to sell it. If poor peasants and landless workers do not have the financial capability, the State can purchase land directly from the landowners with full compensation at current market price, and sell it to them against payment in installments over a long period of time. The alternative is the agricultural credit market mechanism.

The anti-state intervention donors and the World Bank-cum-IMF tend not to provide a number of options but a single blueprint. They virtually insist that financing agriculture and the supply of inputs for agricultural production must be done through a dominant private sector, at the market rate. Progressive land taxation is undesirable, as it is considered to be harmful to landowners' incentives and reduces savings needed for technical progress. In rural areas like the rest, involuntary unemployment and poverty are to be alleviated by encouraging foreign and domestic private investments to raise labor productivity *(USAID*, 1986 and *World Development Report,* 1983, Part II and 1990). In contrast to government intervention, they claim that this course would make markets and incentives work in agriculture, and enhance growth prospects.

It is true that markets and profit-making incentives can work in North African countries. The market works, where equal opportunities and a firm government commitment to prevent a rise in poverty exist. It is also true that empirical evidence suggests that the Western conception of a free exchange of land property and means of production does not even exist in many rural areas. Land has a high amenity value and an intra-family heritable bond that does not always make it a tradeable commodity. Also, it is most likely that where the distribution of land property is highly skewed, the market works for the benefit of landowners, multinationals and traders, while most probably harming the poor peasants and the powerless landless workers. Moreover, how can small tenants and wage- dependent landless workers respond to the market price incentives and profit opportunities from technical change if they have neither secure tenure rights in land, nor legitimate access to land ownership and credit?

The reader may better wait for the empirical evidence in Chapters 4, 5 and 6 in order to judge whether the pace and forms of change in the land market structure of countries studied would have occurred, in the absence of government intervention in the land tenure systems. Also, the reader will be able to judge that the roles of the State and the market (private sector) cannot be viewed in a contrasting form. Their roles have to be seen in a broader context of inter-relationships, within a particular agrarian situation and specific form of government intervention. There are bound to be market failure and government failure, and we need to examine what accounts for the success or failure.

The Policy-making Complexity

In North Africa, state intervention to adjust landed property rights, and to limit the economic freedom of agricultural producers, creditors and traders has two dimensions: political and Islamic-based morality. Though interconnected, they are described separately, and a discussion of their coalescence follows.

The Political Dimension

Like in other parts of the world, the content of land reform-based rural development strategy in North Africa is chiefly a political issue. Questions on public interest in private land, the choice of the policy content and who benefits, are complex and have no definite answers. They have to be examined within the political economy of each country.

Vested Interests. In practice, who gains how much and who loses how much, is a result of conflicting vested interests within the occupational groups of rural people. The conflict of interests is among landowners, tenants, share-croppers, tribal nomadic or agro-pastoral people, traders, intermediaries, and employed farm managers, on the one hand, and between them and the State, representing the public interest, on the other. The former groups, in search for their best interests do not always perform in the public interest of the society at large. Land, soil and water conservation from one generation to the next, social stability, and the prevalence of justice are a few illustrations of public interest guarded by the State through legal procedures. However, state institutions and government employees cannot be considered neutral, representing and safeguarding public interests. Nor are they to be understood as Plato's puritan 'Guardians' endowed with high moral standards and meticulous ethics. Within state institutions, there are elements of vested interests, class bias, and probably corruption. This is particularly so where the supply of credit is administered by public agencies, and large areas of cultivable land are owned and managed by the State. In distributing these lands, the policy makers -- together with the bureaucratic establishment -- are likely to be in favor of particular classes on whom the government depends for its tenure in office.

Lobbying Structure. Where large owners of land and capital in agriculture are few in number and their economic power is dominant, their interests tend to carry heavy weight in policy formation, and in levying land tax. Thus, the lobby of a few can swing the balance of political power to the disadvantage of the large numbers of the *fellaheen* and nomadic pastoralists who have no lobbying strength and no means for collective action. They neither know how to lobby, nor how the political system of their countries is organized.

It follows that given the administrative capability of the State, the choice of land policy and the design of rural development strategy mirror the vested interests and their power bases, including those of who run the country. Together, they constitute the configuration of political power that makes policy. The outcome of such process can take diverse forms. It can take the form of no, high, or low ceilings on privately-owned land, its nationalization or payment of full, part or no compensation to affected landowners. It can include the prohibition of tenancy arrangements or control of farmland rental value, and the choice between prohibiting hired labor and fixing a minimum wage rate for hired agricultural workers, or to leave wages to be determined by the labor market mechanism. Moreover, it comprizes the permission or prohibition of private investors to encroach upon the nomads' grazing lands for commercial farming. The configuration of political power can also determine the disposition of state-owned land to the landless farmers or to large landowners, civil servants, members of the armed forces and multinational companies. It determines also the choice of sites of irrigation canals, rural roads, schools and health centers in the countryside.

The Policy Content. The above paragraphs suggest that the content of public course of action is subject to a combination of the following:

1. The extent of the government's firm commitment to realize social justice and maximize net social gains for the benefit of the poor *fellaheen.*
2. The country leadership's calculus of the relative weight of the power bases of those who hold power. They include: holders of economic power in agriculture and in trade; the moral power of conventional religious leaders and popular strength of Islamic fundamentalists; and the pressure exercised by non-governmental institutions such as the free press, farmers' unions, women's groups and organized syndicates of professionals, including the government employees. Last but not least, there is the power of the armed forces establishment.
3. The form of government deciding on land policy, whether by a parliamentary majority or by authoritative and autonomous power.
4. The existing rural development problems to be tackled, e.g. the extent of landlessness, hunger and conditions of rural poverty, and the size distribution of land ownership together with other productive assets in agriculture.

These suggested elements are tentative and the order of their listing does not imply their relative weights. Despite the progress made in welfare economics, the theory of the State and the theory of collective action, the quantification of the relative weights of all factors influencing policy is still in its infancy. Recently, serious attempts have been made by a number of scholars (Van der Hoeven, 1981, and de Janvry and Sandoulet, 1989). Van der Hoeven constructed an index of power consisting of the number and size of groups of people; their respective average income and education level -- all are relative to the total population. De Janvry and Sadoulet, on the other hand, constructed a political feasibility index and tested it. Their method is plausible with respect to assessing quantitatively an *ex-post* redistributive policy impact on net national (social) gains in terms of increased gross national product (GNP) and the rise in average real income among the socio-economic groups in the rural sector (1989, pp. 18-22 and Table 2).

Nevertheless, economic power is not the only influencing factor. There are other complex elements in the making of rural development policy, which seem to be beyond professional modelling capability. They rest on value judgement. Apart from ethics, examples of pressure groups having *no economic* power do exist in North Africa. They have influenced (some even made) land policies and rural development path. These groups include a small group of both army officers and leaders of independence movement, Islamic fundamentalists, and committed individuals of the intelligentsia.

The Islamic Moral Dimension

The content and manner of the application of Islamic principles governing the moral foundation of property rights, income transfer, savings, profit making, and abolition of exploitation in economic transactions are complex.

Some Moral Principles. As we understand the moral principles which are relevant to our study, they are sketched as follows:

1. The institution of private property in land and capital is guaranteed by Islam provided that it does not violate the Islamic goals of justice and social welfare *'masāleh'*, and that all parties concerned are treated equally by Law. Endowment of landed property and water are to be used for the welfare of all in the Moslem community without exploitation of the weak, orphans and needy, and without bribery, speculation and hoarding.[1]

2. The imposition of giving *Zakāt*, i.e. alms as a portion of earnings above a minimum level *(Nisāb)* specified in detail in the *Shari'ah*. This form of a compulsory transfer of income is for the benefit of the poor, orphans and needy.[2]

3. The strict prohibition of charging interest *(riba)* in lending money. It should be replaced by a partnership in transaction between the supplier and the user of capital on one hand, and the bank as an intermediary charging administrative cost, on the other. Based on arrangements agreed upon in advance, the three parties share in both profit and loss *(Moshārakah).*[3] In capital transaction sense, this procedure amounts to an amalgamation of profit and interest, risk bearing and capital transaction (see a discussion of this issue in Chapter 6, Sudan).

4. Islamic *Shari'ah* courts of law are to settle conflicts in human conduct and to ensure the observance of the rules governing inheritance. Also, they deal with the disposition of one's landed property after death for the benefit of heirs *(waqf ahli)*or for Islamic charitable purposes *(waqf or habous khairy).*

Directly or indirectly, this simplified framework of principles tends to have an important bearing on policy formulation, private savings and the supply of agricultural credit. The degree of their influence depends upon the extent of state instititions' commitment to, and the actual implementation of, these principles.

Likewise, observance of these principles depends on the Moslem individuals in practising Islamic teachings in their daily transactions.

Policy Implications. The question is: given these complex considerations, to what extent should the secular State use its authority and sovereign power to limit private property rights and, in turn, the economic freedom of the owners of land and capital. Should these rights be preserved on the grounds that they are sacrosanct in the worldly capitalist thinking? Or, should the violated private property in land and other means of production be abolished (nationalized) in the worldly socialist system of political thought? Should the Moslem community base their economic behaviour on the following principles?

1. *Allāh* (God) guides every Moslem through the *Qur'ān, to resolve interpersonal conflicts and to follow a straight path (Al-Sirāt al-mostaqeem)* about which each individual is accountable to Him after death on the Day of Judgement;
2. The legitimate rights in ownership and use of land and water by His creatures are to be protected and directed for the interest *(Masāleh)* of the Moslem community; and
3. The welfare of the poor and needy conditions the total welfare of the Moslem society.

In short, the moral force of Islamic principles requires that in both secular and non-secular states, a few should not benefit while many others are harmed and live in misery. The entitlements of the poor *(foqarā', yatāma* and *massākeen)* and the responsibility of the Moslem society towards the poor are stated in 36 verses *(Ayāt)* appearing in 23 chapters *(Soura)* of the *Qur'ān.*

The Meaning of Exploitation in Economics and Islamic Principles

The terms 'exploitation' and 'exploitative relations' in production and exchange have been frequently used in statements by the North African governments justifying their intervention in both the land and credit markets.

Though commonly used, the term 'exploitation' is ambiguous and surrounded by contention. It does not, therefore, possess a settled meaning on which all users agree. Yet, its usage has significant policy implications for human conduct and transactions associated with 'other people', 'things' (e.g., land, capital), or 'institutions' (e.g. government, co-operatives, banks). The anti-exploitation policy implications include expropriation with payment of compensation, or confiscation of property, and the limitation of economic freedom of the private sector, including administered rental value of land. They also include the abolition of *riba* in banking transactions. The prevention of exploitation seems also to be behind the state monopoly in the supply of means of production and its procurement of harvests at fixed prices.

How is exploitation idiomatically expressed in words, and operationally viewed by governments? In what way is it academically measured by economists and morally perceived by Islamic principles? According to Webster and Oxford English dictionaries, the idiom of exploitation is phrased as "the act of exploiting", "the action of utilizing for selfish purposes", "selfish and unfair utilization", "making use of basely for one's own advantage or profit", "taking advantage of . . .". Thus these definitions diagnose exploitation with a moral emphasis on undesirable human conduct of business.

The Operational Meaning

Operationally, governments, in their statements accompanying land reform legislation and laws prohibiting *riba*, view exploitation as a mix of illegitimate profit-making and a violation of economic freedom in a capitalist agrarian system. Often, they equate pre-reform systems with feudalism *(Iqtā')*, the term borrowed from pre-Seventeenth Century Europe. Feudalism in its *real* sense (with coercion of serfs and military service to landlords) did not exist in the immediate pre-reform period throughout North Africa. Nevertheless, the term was commonly used by post-independence governments in justifying their intervention. They have frequently expressed the presence of exploitation in terms of the following socially harmful and undesirable institutional arrangements:

1. Landownership concentration.
2. Landlords exacting illegal levies and services from their peasants.
3. The dispossession of the natives' lands by foreigners during pre-independence period.
4. Large land owners' and merchants' collusion with the former government administration to depress the earnings of tenants and landless workers.
5. Widespread absenteeism among private landowners, separating ownership from management and investment associated with (a) renting-out land at exorbitant rental rates under unwritten and insecure tenancy, (b) no compensation payment for eviction or improvements, and (c) persistent accumulation of wealth and privileges contrasted with the wretched life of the *fellaheen*.
6. Prevalence of money lending at usury rates of interest, leading to indebtedness which eventually results in loss of land to creditors.
7. Illegal extortion of land *(Ghasb)* by virtue of official status and political power.

Academic Meaning and Measurement

In academic debate, the controversy over the meaning of exploitation was initiated by Karl Marx's conception of transactions associated with private ownership and exchange of the means of production in a capitalist economy. He explained the mode of market power relations in the use and exchange of resources which could be viewed as exploitation. In his intellectual construction, Marx made it necessary to view the separation of society into an exploiting and an exploited class as a consequence of capitalist mode of production.

Based on his compiled data from Irish and English agriculture during the period 1851-1871, Karl Marx conceived exploitation in terms of the capitalist's extraction and accumulation of 'surplus value', particularly from displacing labor by newly invented machinery of his time, and from setting wages at subsistence level. He considered the volume and rate of appropriating surplus value as a measure of exploitation. From the class conflicts within capitalist agriculture, he identified the owners of land and capital as the exploiters, (e.g., landlords and the spinners of cotton and wool), and the wage-earners (laborers), as the exploited living in 'increasing misery'. The prevailing exploitative relations in production, in his words '. . . have sprung up historically and stamp the laborer as the direct means of creating surplus-value.' (*The Capital*, part V, p. 558).

Another group of exploiters was added by Lenin in his study of late nineteenth century Russian rural economy. In that work, 'The Development of Capitalism in Russia' (1899), Lenin considered the middle farmers with a commercial orientation of their means of production as exploiters of those who own only their labor power and working animals. But his generality lacked the identification of the exploitation *criteria* in production and exchange which were identified during the 1920s by a group of Russian researchers. Based on field surveys, they arbitrarily allocated points for different agrarian transactions. These were primarily hiring labor, renting land, leasing out draught animals, and dependence of hired workers on others for the sale of their labor power (Cox, 1986: Tables 5.1 -- 5.11).

An alternative academic meaning of exploitation comes from two American scholars, Robert Nozick (1974) and John Roemar (1982). They go beyond the Marxian hypothesis, both attacking his explanation which they consider applicable to pre-capitalist feudal conditions. From their point of view, exploitation is seen as a violation of property rights and entitlements. Each places a different emphasis in his system of analysis. Nozick argues that the exploited do not possess the scarce entrepreneurial abilities and skills to innovate and to bear market risks together with uncertainties in free market transactions. Thus, his argument is extracted from a competitive private property-market economy.

For Roemar, the key to exploitation lay in the availability of alternatives and freedom of choice. Workers are exploited, when they cannot form a coalition (such as a trade union) in order to withdraw freely from the exploiter to become better off, whereas their exploiter becomes worse off. Another criterion is the inequality of opportunities manifested in an uneven access to means of production other than labor because of imperfect functioning of the competitive market. Under socialism, on the other hand, access to the means of production including land is, in principle, secured to all farmers. Exploitation in socialist agriculture takes the form of limited freedom of choice and inequalities, due to relationships of dominance by Party members having privileges granted to a few in a controlled economy.

All these examples of academic contributions to the meaning of exploitation are formed by individual scholars with varying backgrounds, and within each author's unique system of analysis. Abstracted from different conditions of production structure and systems of remuneration, the analytical reasoning behind each interpretation seems to be consistent with each conclusion reached. It appears also that the criteria employed in measurement are based on value judgements, having ethical implications for the notion of inequality and social welfare. This brings us to the moral connotation of exploitation in the Islamic principles.

The Moral Meaning in Islam

In their conduct, Moslems' consciousness of their mutual obligations in practising what is permitted *(halāl)*, and refraining from what is prohibited *(harām)* in the *Qur'ān*, is essential. They are also required to perform with a high standard of honesty in financial transactions and trade. Foremost in the range of prohibited human activities is usury *(ribā)* combined with dishonest or fraudulent transactions. The *Qur'ān* teaches us "Those who live on usury shall rise up before *Allāh*, like men whom Satan has made evil by his touch, for they claim that usury is like trade. But *Allāh* permitted trading and prohibited usury" *(Sourat al- Baqara:* 275).

Islam prohibits exploitation in the form of usury, transgression against the property of others and charging tenants 'unfair' rent or evicting them without fair compensation. Prohibited also is the violation of the poor and orphans' entitlements. Nevertheless, the interpretation of Islamic principles about these relationships and their application to concrete situations has continued to be a subject of controversy.[4] For example, are all interest rates act of exploitation? What is the 'fair' or 'unfair' rent? Are central banks or agricultural banks, charging interest on loans and paying it for savers, exploiters?

While there is a consensus on the prohibition of *Riba* as the cardinal sin, there are diverse interpretations of modern banking transactions with regard to exploitation, not only between the major sects *(Sunni* and *Shi'ah),* but also within each sect *(Madhāheb* and *Toroq).* Although different intepretations exist, their principles are derived from one fundamental source, the *Qur'ān* and the explanation of its fundamental rules by the Prophet saying *(Hadith)*. For example, whereas some Islamic jurists have considered 'low' interest rate permissible, and 'fair' charges are legitimate, others strongly argue that charging *any* interest for transaction is usury *'Riba'*, and that it should be abolished and replaced by partnership in sharing benefits and loss *(Qard hassan, Moshārakah, Muqāradah)*. The latter is considered not a form of exploitation but a permissible dealing (see note 3). The holders of this view on permissible business relationship call for governments to officially institute these arrangements through central banks (Uzair, in Ahmad, 1981 and Tantawy, 1992).

Concluding Remarks

The above discussion on the interpretations of exploitation by governments, economists and Islamic jurists suggests the continuing ambiguity of the term and its equation with unjust transfer of income from a poorer to a richer person leading to widening inequality of personal welfares. However, the technical sense in which the term is used manifests different treatment in the professional literature. Marx's appropriated surplus value corresponds, in part, to the classical 'economic rent' or monopoly profit resulting from the monopolist's deliberate wage setting below the value of the worker's marginal product in an imperfect competitive market. Appropriation of monopoly profit is also an approximation of Islam's moral notions of preventing unjust and deceptive dealings with persons and things. Yet, the academic debate continues and the conflicts between governments and Islamic fundamentalists (supported by a group of active writers on Islamic economics) is steadily increasing.

Food Security

Being a normative concept, food security is a controversial concept. The term may be generally applied as the counter meaning to food insecurity, which in the Oxford Dictionary denotes "not safe and dependent". It follows that a food-insecure household is just as insecure as a food-deficit country which depends upon donors' aid and imports to satisfy its food necessities. Such dependency is susceptible to changing international political relations (e.g., the threat of using food as a political weapon) and world trade regime. However, our concern is more with the rural household and individual consumer than with the country as a whole. Within this concern, special emphasis is placed on the linkages between legally-secured access to land, food security (FS hereafter) and the process of rural development. FS is viewed here as the secure access to nutritionally adequate and stable flow of food supply by rural people in general, and the poor in particular.[5] It follows that a greater self-sufficiency in food production promotes food security.

Sources and Measurement of Food Insecurity

Understanding FS in North Africa requires, therefore, the identification of both the standard boundary between it and food insecurity, and the sources of fear and threat of the latter. The possible sources include:

1. Uncertainty of rainfall and prolonged drought.
2. Crop failure.
3. Low economic capacity (meagre assets owned and low wage earned) to aquire the needed amount of food.
4. Instability of food aid due to changes in political relations between donor and recipient countries.
5. Sudden rise in grain price caused by a coalition among traders.
6. Civil unrest or war.

With regard to the identification of the causality of food insecurity, it is a matter of value judgement. Likewise, the establishment of a measurable standard (cut-off point) of adequate *per capita* daily supply of calorie and protein is arbitrary. Nevertheless, there are private and public concerns about estimating how far and how many who fall below such standard of food security. The estimates are, therefore, useful in suggesting the scale of the country's food self-sufficiency. Estimation is also needed to form a judgement on the incidence of undernutrition or hunger and related risk of ill health among the consumers by sex, age, occupation and by localities (rural and urban).

The results of household surveys on the actual food consumption provide indications, having policy implications. Given the diverse systems of social organization of agriculture as well as the climatic conditions of each country, policies vary. They range from tackling the root causes to dealing with the consequences of food insecurity and hunger. They also vary from policies aiming to benefit population at large, to programs targeted to benefit the frequently vulnerable sections of the rural population. The benefits from the former are likely not to reach the rural poor and certainly not the poorest due to leakages in implementation by the bureaucracy. In both approaches, it is essential to increase the production of food and to ensure the adequate and stable flow of food supply in order to gradually eliminate staple food imports and the country's reliance on the politically vulnerable food aid.

Food Insecurity and Poverty

Because food is a biological necessity for human survival, the adequacy (or inadequacy) of access to food is closely associated with changes in poverty level. The food insecure people who are undernourished or unable to acquire the minimum daily calorie level per person are almost always poor. They tend to suffer from low capabilities in their rural localities. The converse need not be true. The poor are not all hungry or undernourished. As one would expect, and as will be verified from available empirical evidence in the rest of the book, the chief determinant of this association is the low asset ownership, together with low and unstable flow of earnings, viz., low effective demand. Our proposition is that a legally secured access to productive land is a household life asset, crucial for providing it with FS and is an insurance against the risk of undernutrition and absolute poverty. This is to be realized by way of a combination of intensive family labor use to raise productivity, increasing self-produced food, and the pricing of crops. This proposition is likely to be valid in agrarian economies where alternative and stable employment opportunities are limited and in periods without natural calamities.

Technically, there is no reason why a wider access to land and credit should not augment food crops' yields and realize the small farmers' food security. There is also no reason why a macroeconomic price policy should not induce small farmers to produce food beyond their consumption necessities and at the same time produce non-food crops to acquire economic surplus which benefits the national economy.

We know from the ideas behind Engels' law (1856) and the economic principle of marginal propensity to consume, that a rise in the small farmers' income leads to a fall in their income elasticity of demand for food, and a reciprocal rise in the demand for non- food items (fuel, clothes, furniture), which are normally domestic products. Thus, with the multiplier effect of ensuring FS combined with increasing farm production, savings of small farmers are likely to rise and the employment of labor in the domestic economy expands. Concomitantly, the rise in the productivity of the rural poor and the improvement in their nutrition tend to break the vicious circle of food insecurity: low food-intake leads to low physical activity and productivity, which in turn leads to low income, low purchasing power and low food-intake. These linkages leads the discussion to the conception of rural development.

On Rural Development

The conceptual issues discussed so far coalesce in one meaning: the rapid alleviation of undernutrition and absolute poverty which is synonymous with the prime objective of rural development. The aim is the enhancement of the productive capabilities of the rural population, particularly the poor and the realization of their food security, together with the economic potentialities of the scarce agricultural land and water.

The Meaning of Rural Development

In policy design and ordinary parlance, rural development has a more ambiguous connotation than do land policy, exploitation and food security.

Analytically, it is more difficult because of its several constituents. Often 'rural development' has meant the construction of schools, clinics, youth clubs and roads as well as the expansion of non-farm jobs in selected rural areas. The improvement of land tenure arrangements and the realization of food security and equity in the benefits from augmenting agricultural output have usually been either overlooked or underestimated.

In the phrase 'rural development', the confusion seems to lie in the sense in which the word *'development'* is used. If it refers to *things* like the provision of irrigation water, chemical fertilizers, and the construction of rural roads, it cannot have the sufficient impetus of development. Though necessary, these 'things' are insufficient without improving income distribution, ensuring access to adequate and stable flow of food supply, and without raising average real income per head of the rural poor. If it refers to *people,* then one asks whether it refers to *all* people, rich and poor, residing in rural areas, or to the poor and poorest as the beneficiary target. Thus it is the *policy* on rural development or the approach taken to tackle the problems of underdevelopment that we judge.

Logically speaking, I think, the force of the word 'development' lies in its contrast to the term 'backwardness' or underdevelopment. Rural development strategy choice, therefore, must tackle the causes of underdevelopment either in all the countryside or in certain backward regions to bring about a balanced development in the rural sector, and eventually between it and the urban-manufacturing sector (see Chapter 8). Thus it is basically an anti-poverty strategy, designing the pattern of agricultural growth, providing greater access to cultivable land and complementary inputs, and improving income distribution in the rural sector. The reasoning behind this policy mix is that the present generation of poor *fellaheen* and nomadic people cannot be significantly better off without raising their real incomes and food intake the whole year.

Policies and Programs

Like national development, there are many strategies for rural development. Each has a different policy mix and diverse patterns of agricultural output growth and distributive consequences. Each strategy requires a categorization of problems to be tackled and classes of people to benefit. If the strong preference is the rapid elimination of household food insecurity and absolute poverty in rural areas, the *fellaheen* and pastoral nomads should be the primary beneficiaries. Their problems and expectations are fundamentally different from the rest of the rural population.

Rural development is, therefore, a serious and enormous task. Being a dynamic process, the benefits accrued do not come from a once-and-for-all policy instrument. They are brought about in a sequence of changes pursued consistently by persistent efforts of government and non-governmental organizations. The cumulative effect is a sustainable rise in the real income of the poor and the poorest at a rate faster than that of the rich. The long-run effect is a balanced development beween the rural-agricultural sector and the other sectors of the economy.

It seems that rural development principles are one thing and practice is another. In practice, rural development (its content, concrete function, and its potential distributive benefits) are differently viewed. They are seldom agreed by government programers, donor countries, and international assistance agencies. What they seem to have in common is the broad phrase "raising the level of living of rural people". As a consequence of this aggregation, and for reasons of land-based power structure and limited financial means, the poor and the poorest, in particular, tend to be by-passed. Past experience suggests that in a locality-specific rural development project, the allocation of foreign and domestic financial, political and technical resources tend to be disproportionately captured by a few influential landowners and traders. In such situations, the rural development foreign prescribers find it safer to avoid the 'politically sensitive' issues of land tenure. They instead, prefer to focus on the merits of delivery of services. Consequently, the emphasis in the evaluation of these projects is chiefly placed on a rather administrative approach i.e. the realized disbursement of materials and delivery of services versus what was planned.[6]

We have argued, elsewhere, that the reduction of rural poverty incidence both proportionately and in absolute numbers is *the principal operative criterion for judging rural development* (El-Ghonemy, 1990: pp. 91-3 and chapter 7). This is to be chiefly realized by greater access to productive assets, notably land and education, and the resultant improvement in nutrition and abilities of the *fellaheen*. Together, they interact with the following structural forces operating in the national economy.

1. The adequacy of investment in human capital (health, education) and in augmenting the aggregate supply of irrigated land and its productivity.
2. The quantum and quality of trained technical manpower required for rural development programs.
3. The demand for labor outside agriculture and the resulting reduction of density of labor per unit of land.
4. Factors affecting the level of demand for agricultural products in the domestic and international markets.

Collectively, and in a private property-market economy, the interplay of these supply and demand forces with the size distribution of land regulate the distribution of income (and consumption) in rural areas. Specifically, and depending on the original distribution of income among the rural poor, the growth rates of output and rural population determine the rate of increase in the average income of the poor and the speed of their crossing the poverty line established by each country at a given point in time.

In short, the reduction of poverty and undernutrition represents the point where all these structural forces converge *via* improvement in income distribution (see Figure 1.1). We shall have the occasion to go into these relationships a bit more in chapter 7 when we examine the empirical evidence of changes overtime in technological change, land concentration, agricultural growth and rural poverty characteristics.

Notes

1. For a comprehensive analysis of these conflicts, see chapter 7 in Khurshid Ahmad (editor), 1981. See also Ibn Khaldoun, *Al-Moqaddema* explaining the political philosophy of Islam (English translation, 1967), and Kedourie's article in The Daily Telegraph, 18 June 1990, London, on the reasons for the Islamic fundamentalists' victory in Algerian election.

2. There is a rich literature on the meaning of *Zakat* and its payment as well as its role in modern fiscal policy. See a list of references compiled by the International Centre for Research in Islamic Economics, King Abdul Aziz University at Riyadh, Saudi Arabia and published by the Islamic Foundation, Leicester, U.K., edited by Ahmad, 1981. These references appear in pages 290-2 and 310 of the above publication.

3. Since the mid-1970s, and at the initiative of Prince Faisal of Saudi Arabia, these principles have been applied to lending procedures, savings and investment in a network of branches of Faisal Islamic Bank established in most of the North African Countries. Some other commercial banks have also changed their operations to conform with these principles. The principle of *Mushārakah* or partnership requires that the bank and the client contribute to an agreed investment operation. Both parties agree to contribute to the capital and to divide the net profit in a proportion agreed upon in advance. For a detailed account of this type of banking procedures, see *Faisal Islamic Bank*, (1979) and (1983). Also see a survey of literature on *Riba*, interest-free banking, and profit sharing in pages 287, 296, 306, 310 of the publication cited above in Note 2 and by the official Mufti of Egypt, Mohamad Sayed Tantawy, *al-Halāl wal-Harām Fi Mo'amalat al-Bonouk*, published by *al-Ahram al-Iqtisadi*, 20 January 1992, Cairo. On the application of these principles in the Sudan, see Shaaeldin and Brown, chapter 8, in Barnett and Abdel-Karim (editors), 1988. On the capitalist elements in these principles, see Gibb (1950).

4. On the different interpretations of the classical Islamic rules on tenancy and land taxation, see: in Arabic, among many others, Abu Yusuf, *'Kitāb al-Kharaj,* 1382 Higri'*, al- Qadouri *'al-Kitāb',* al-Sarakhsi *'Kitāb al- Mabsoot',* and al-Shaibani *'Kitab al-Asl'.* For a review and analysis of the literature on this subject by a non-Arab scholar, see Johansen, *"The Islamic Law on Land Tax and Rent"*, Croom Helm, 1988.

5. Since the World Food Conference (Rome, 1974) the subject of food security has progressively been the focus of the work of the Food and Agriculture Organization of the United Nations (FAO). In 1984, its governing bodies approved a broad definition, "ensuring production of adequate food supplies; maximizing stability in the flow of supplies; and securing access to available supplies on the part of those who need them", (Committee on World Food Security, nineth session, Rome, 11-18 April 1984, CFS: 84/4, p.2). Another definition is provided by the World Bank. It defines food security as 'access

by all people at all times to enough food for an active, healthy life', *Poverty and Hunger: Issues and Options for Food Security in the Developing Countries*, 1986. On measurement problems, see John Staatz *'Measuring Food Security in Africa: Conceptual, Empirical, and Policy Issues'*, *American Journal of Agricultural Economics*, December, 1990.

6. See my assessment of rural development programs and projects in El-Ghonemy (1979, pp. 5-7) and in Yemen, UN/ESCWA, 1986. For a critical review of rural development projects in Africa, see Uma Lele (1979). For a careful review of integrated rural development projects in several countries, and the bureaucratic bias against rural people, see Ruttan (1984).

3

A Regional Perspective of Rural Development Problems

This problem-focused chapter examines the structural and institutional characteristics of North African agriculture and related rural development policy issues. It provides the background which helps the reader to understand the nature of inter-country variations in agricultural population density on land, and food security problems. Also, it explores briefly some aspects of historical and institutional importance which help to understand the current social organization of agriculture. Issues on ownership and allocation of agricultural land between food and non-food crops, and on the supply of complementary inputs are discussed in the following three chapters.

Introduction

Apart from being ruled by Europeans for a considerable period of time, North African countries have common cultural features: Arabic, Islam and customs. Most generalizations conceal considerable structural differences in: natural resource endowment; size of population; proportion of nomadic people; average *per capita* income; the relative importance of agriculture in the economy; and the degree of food self-sufficiency. Within agriculture, there are also great variations in: density of working people on land; the area of potentially cultivable land; and the extent of reliance on rainfall for staple food production. Even in common cultural features, the countries differ in rural women's equal rights, particularly with regard to access to land and membership of agricultural co-operatives. Countries also differ in the strict application of Islamic principles related to economic transactions, gross inequalities and property rights, including the Islamic land tenure arrangements of *Waqf* or *Habous*.

TABLE 3.1 Some Structural Indicators of North African
          Countries, 1960-1990

Indicators		Algeria	Egypt	Libya	Morocco	Sudan	Tunisia
1. Total population, million	(1990)	25.1	52.2	4.5	25.0	25.1	8.1
-average annual growth	(1960-1970	2.4	2.5	3.9	2.6	2.2	2.0
rate %	(1980-1988)	3.1	2.6	4.3	2.7	3.1	2.5
2. Agricultural population							
as percentage of total	(1989)	24.5	41	14.1	37.2	61.4	25.2
3. Rural population							
as % of total	(1987)	56	52	33	53	79	46
- estimated nomadic pop.							
as % of rural pop.	(1976-1980)	3.0	0.5	10.0	17.0	15.0	2.0
4. Gross National Product, US$							
per person	(1988)	2382	660	5420	830	480	1230
5. Agric. GDP							
- as % of total GDP	(1988)	13	21	4(1985)	17	33	14
- average annual							
rate of growth %	(1965-1980)	5.8	2.8	10.7	2.2	2.9	5.5
	(1980-1988)	5.6	2.6	7.3 (80-85)	6.6	2.7	2.4
- direct contribution of agric. GDP growth to total GDP growth %		20.9	9.6	-	26.7	35.6	9.9
6. Agric. GDP per capita labor force (US$ at 1980 prices),							
annual growth rate %	(1970-1980)	7.9	2.2	n.a	1.2	-0.1	3.5
	(1980-1990)	3.2	1.2	n.a	5.5	0.9	2.9
7. Exports:							
- agric. exports as % of total	(1987)	1	17	1	16	97	9
- oil & minerals as % of total	(1987)	97	74	99	26	6	27

Sources: Rows 1, 3, 4, 5 and 7: *World Development Report* 1987, 1989 and 1990: Development indicators. Rural population: calculated as the balance of urban population. Estimates of pastoral-nomadic population are rough; they are taken from countries population censuses and Ian Livingstone (1984: Table 2, columns 4 and 7) except Sudan: Ministry of Agriculture, 1978.
Direct contribution of agric. GDP growth is for the period 1980-88, calculated by multiplying the rate of growth of agric. GDP and its percentage share in total GDP divided by the rate of growth of total GDP.
Row 2: Calculated from *1990 Country Tables - Basic data on the agricultural sector*, FAO, Rome.
Row 6: World Bank and FAO Data Base.

The region comprizes countries of different types of economic structure. They range from a predominantly agricultural low-income economy (Sudan) to middle-income economies (Algeria, Egypt, Morocco and Tunisia). The region also includes an oil- based and high-income economy (Libya). The variation is so wide that, in 1989, the average annual income per person in Egypt was one-tenth of that in Libya and the latter was more than 12 times the average in Sudan. As we shall soon find out, the region comprizes land-abundant and labor-scarce economy, Libya, land-and labor-abundant economy, Sudan, and a land-scarce economy, Egypt. Tables 3.1 and 3.2 reveal two striking, but related features. One is that all countries are dependent for meeting their food requirements on imports, and their self-sufficiency rate of cereals fell in the 1980s relative to the 1960s. The other is the accelerated growth of population, being overwhelmingly rural, at the average annual rate of 3.1 percent which is higher than the average rates in Sub-Saharan Africa (2.7%), Asia (2.3%), and Latin America (2.2%) in the 1980s.

Although the importance of agriculture relative to other sectors of the economy has declined over the last three decades, yet it provides employment for a considerable number of the agricultural labor force (except Libya). As computed in Table 3.1, growth of agricultural GDP in the 1980s was responsible for more than one third of Sudan's total GDP growth, 26.7 percent in Morocco, 21 percent in Algeria and about 9 percent in both Egypt and Tunisia. Compared to the period 1965-1980, the slowdown in agricultural growth rates in the 1980s has held back national economic growth.

An Historical Brief on the Origin of Duality in Agriculture

The independence of the six countries of North Africa, as State nations, is relatively recent, mostly since the 1950s.[1] All were ruled by European countries for a considerable period of time; the longest colonial rule being in Algeria. During their rule, the Ottoman Turks, the French, the British and the Italians imposed their own legal systems and forms of government upon the indigenous systems of social organization. They also linked the agricultural policy pursued in the colonized countries to their own home policies, in order to serve their primary interests.

The historical experience of the countries suggest that, starting in the sixteenth century and for a period of nearly three centuries, the prime interest of the Ottoman rulers was fiscal. They extracted exorbitant land taxes from the *fellaheen*, through assigned local agents in exchange for large landholding grants. Subsequently, the French and Italian rulers in the Maghreb were chiefly concerned with colonizing vast areas of the best agricultural land to settle their own individual and corporate farmers. The 'colons', using their technical knowledge and native cheap labor grew wheat, sugar beets, and vineyards for wine-making against the Islamic principles which prohibit drinking alcohol. On the other hand, the British Administration in Egypt and Sudan did not colonize lands for British farmers. Instead, the interest was in expanding cotton production and its export to the British textile industry. Another concern was the promotion of the British private sector's involvement in land development, and commercial activities. Hence, the irrigation expansion in the Nile Valley.

Nevertheless, the French, Italian and British colonial rule introduced elements of agricultural modernization, and land and credit markets: purchase, sale, charging interest for lending and the introduction of contractual arrangements for leasing and mortgaging land. They have also increased the aggregate supply of cultivable land in the colonized countries through technological change.

Property and usufruct rights in land were also granted by the colonial administration to certain tribal chiefs, senior officials and influential native families, on whom the power of the foreign rulers depended. Both categories of large farmers (foreign settlers and native families) held a dominant power in the land market structure. The source of foreign economic power in the rural economy of the colonized countries is revealed by official records at the time of independence. The area of foreign-owned lands (individuals and land companies), as a percentage of total cultivated lands was 30 in Algeria, 20 in Tunisia, 11 in Libya, 6 in Egypt and 11 in Morocco. In Sudan, the Gezira scheme was not owned by the British. It was a triple partnership between the Sudanese government, the British management and the cultivators. Totalling 0.2 million hectares in 1956, when Sudan gained independence, the area of the scheme was fully irrigated, growing mostly cotton, and it represented 30 percent of total irrigated land, but only 2 percent of total arable land.[2]

Hence the formation of large estates was initiated and the duality in the production structure between modern and traditional sectors established. The duality was manifested in the juxtaposition of export-oriented, foreign and large native farms mostly irrigated on the one hand, and the vast area of poorly rainfed subsector, producing at subsistence level and embracing the majority of farming and nomadic population, on the other hand. Thus, a wide disparity in production techniques and wage differentials prevailed, and the process of polarization in the distribution of land and income took root in most countries of the region.

Rural Population: Mobility and Productivity

Out of the region's total population of 140 million in 1990, more than half are concentrated in the Nile Valley (Egypt and Sudan). In the 1980s, the population of all the six North African countries grew faster than during the preceding two decades, mostly due to notable progress achieved in reducing death rates at an impressive pace. Given the slow employment expansion in other sectors of the economy, such a demographic characteristic has implications for: the growth rates of the labor force in agriculture and their productive absorption; their access to land; their *per capita* real income; and their density on agricultural land. Likewise, the fast growing population has important effects on the demand for food. Let us briefly take these effects in order.

With the exception of Libya, working people in agriculture range from one quarter to two-thirds of total. They grew faster in the 1980s than in the 1970s in Algeria, Egypt and Tunisia. Their average annual rates of growth fell slowly in the other three countries, suggesting that alternative employment opportunities in the rest of their respective economies have continued to be limited. In absolute terms, the numbers of working people in agriculture have increased during the second half of the 1980s in all countries, except Tunisia. Alarmingly, annual rate of growth of *per capita* real income fell during the period 1980-90 compared to the period 1965-1980 (except Morocco). Although the usage of averages conceals the size distribution of income within agriculture, this falling real value of productivity must have hit the low income groups harder than rich farmers.

The Pastoral Nomads

A salient feature of the population in North Africa is their mobility in three forms. The first and the oldest is pastoral nomadism on the old-established, communally held land by tribal organizations. Due to their continued mobility in search for rainfall and to their dispersion in vast grazing areas, the pastoral nomads and their livestock are underenumerated in population census, and they are not easily reached by public services. Yet the socio-ecological system of pastoral nomadism employs millions of rural population, and determines the value and quality of production of a considerable part of food (milk and meat), wool and hides. Without pastoral nomadism, a substantial part of arid and semi-arid lands in North Africa would, otherwise, remain unutilized.

As percentage of rural population, pastoral nomads represent approximately 8 in Libya, 15-17 in Sudan and Morocco, 2-3 in Tunisia and Algeria and only 0.5 percent in Egypt. Due to rising conflict of interests between nomads and urban land speculators, together with large commercial farmers who have encroached on pasture land, the nomadic population are gradually restricted within smaller boundaries. Not only have grazing areas been reduced, but they have also less rainfall to support the nomads and their animals.

Rural Outmigration and Rapid Urbanization

The second form of rural population mobility is the steady flow of migration to urban areas seeking higher earnings. Urban population in the six countries has more than doubled between 1960 and 1988; from 28 million to nearly 60 million. About 40-45 percent of this growth is roughly attributed to net rural-urban migration (United Nations, 1980). The centralized bureaucracy, economic activities and political institutions in capital cities and large towns have tended to expand their population. Judging from the post-1960 growth of the largest city in North Africa, Cairo, it appears that the larger the city, the faster its growth (from 3.3 million in 1960 to 6.5 million in 1986, representing 14.5 percent of total population). The United Nations projects that by the year 2000, the population of Cairo will reach 13 million or 21 percent of the projected total. The frequent droughts have also forced many pastoral nomads and small farmers in rainfed areas to migrate to cities, seeking food and jobs.

As urban population do not produce their food, rapid urbanization has compounded the food problems. It has increased the total demand for food, contributed to rising food imports, and changed the pattern of food demand, especially the consumption of wheat, vegetables and meat. Rapid urban expansion has also encroached upon the already limited cultivated land, and raised its price levels which has fueled inflation. For example, the area of the scarce cultivable land in Egypt fell by nearly 420,000 ha. between 1970 and 1986 and the purchase price of land rocketed by 800 percent compared to only 150 percent rise in the crop prices index (El-Ghonemy, 1992).

The third type of rural population mobility is their emigration abroad to Western Europe and to the oil-rich Arab states. Like rural-urban migration, the fundamental cause has been economic, i.e. high wage rate, better quality of life, and wide income gap between native and foreign countries. There was a flow of migrants to France from Algeria followed by Tunisia and Morocco, peaking in the 1970s, then slowed down in the 1980s. Starting in the 1960s there was another wave of emigrants from Tunisia, Egypt and the Sudan to Libya, Saudi Arabia, Iraq and other Gulf countries. This outward mobility has significantly increased after the sudden rise in oil prices in 1974 and 1979. The estimates of the numbers, duration and the composition (rural, urban) of the emigrants are scarce and they vary widely. Scattered information suggests that approximately 15-20 percent of the emigrants were adult male, agricultural workers. Berks and Sinclair (1980, pp. 134-5) estimated that in 1975, the inter-Arab countries' migration from Egypt, Morocco, Tunisia and Sudan was nearly half a million which was almost trebled in the early 1980s before the sudden collapse of the labor market in Iraq, Saudi Arabia and Kuwait, following the sharp fall in oil prices (1982-1986), the Gulf crisis in 1990, and the outbreak of the Gulf War in January 1991.

TABLE 3.2   Agricultural Land, Labor and Food Situation in North Africa 1970-1989

	Sub-Periods	Algeria	Egypt	Libya	Morocco	Sudan	Tunisia
Land[a]							
Total arable, annual rate of growth %	1970-79	0.96	1.76	0.16	0.52	0.58	0.19
	1980-88	0.17	0.66	0.33	1.15	0.09	0.43
Irrigated, annual rate of growth %	1970-79	0.61	1.50	2.54	2.84	0.86	5.65
	1980-88	4.69	0.67	0.81	0.43	0.76	7.59
Irrigated area as % of total arable	1986-89	5.12	97.00	13.30	15.45	15.04	8.30
Agricultural labor Force[a]							
As % of total labor force	1989	25.0	41.0	14.0	37.4	61.4	25.2
Annual rate of change %	1980-89	1.12	1.38	1.06	0.9	1.32	- 0.36
Density, persons/ha. arable land	1985	0.17	2.21	0.07	0.32	0.37	0.14
Food[b]							
Total production, % annual change	1970-80	1.3	1.5	6.2	1.0	3.4	2.1
	1981-88	4.3	4.1	2.5	6.4	1.4	2.7
Per capita productivity change	1970-80	-1.8	-0.9	1.7	-1.3	0.4	-0.2
	1981-88	1.2	1.3	-1.7	3.6	-1.6	0.2
Self-sufficiency rate of cereals including rice[c] (average)	1969-71	74	77	25	95	97	61
	1983-85	33	49	24	67	97	63
Food imports % of total average	1980-86	21.0	29.0	16.0	18.0	19.5	14.2
Food imports dependency ratio[d]	1981	64.5	40.6	69.7	38.2	9.6	42.8
	1988	69.0	47.0	78.0	30.8	12.1	41.6
Cereal aid in 1,000 tons (average)	1984-86	2.6	1874.8	-	330.1	859.0	135.8

Notes: a. Labor force are the economically active agricultural population.
b. Food production annual rates of change refer to the base period 1979-81 (average), and per capita refers to total population. c. is the percentage of domestic cereal production to total domestic human consumption and non-human uses such as animal feed and seeds. d. is the percentage of food imports to total domestic food supply for consumption.

Sources: a & b are computed from *Production Yearbook*, several years and 1990 *Country Tables*, FAO, Rome. c. 'World Agriculture Toward 2000', an FAO study edited by Nikos Alexandratos (London: Belhaven Press, 1988), Table A.5; d. 'Human Development Report 1990' UNDP; and 'Food Aid in Figures', (FAO: Rome, 1987).

The impact of rural outmigration differs not only among countries, but also within rural localities in each country. Unfortunately, micro-studies and quantitative information on this critical matter are either shaky, or do not exist. This is particularly so with regard to the impact on the rural labor market, distribution of income in rural areas, and on the land market. Important among the effects, which have not been adequately researched (except Egypt, IFPRI, 1991), are the allocation of remittances between consumption and direct investment in agriculture, and the extent to which the temporary shortage of emigrating adult male labor induces the use of farm machinery as a substitute. However, household surveys and population censuses suggest that rural outmigration results in agriculture having a less educated and mostly illiterate population with a higher representation of elderly male farmers, and child and women labor. The surveys also suggest the proportionate increase in non-agricultural sources of rural household's income.

Increasing Scarcity of Cultivable Land

In 1990, out of the total area of the six countries, 838 million hectares, only 10.6 percent (about 89 million ha.) are cultivable, 45 percent of which is already under annual and permanent crops. Of total future cultivable land, 76 percent is in Sudan, 10 percent in Libya, 5 percent in Egypt and between 3 and 4 percent in each of Algeria, Morocco and Tunisia (FAO, 1986). A small portion of the potentially cultivable lands is rainfed and most of it must be irrigated.

Agricultural land has been viewed by most people of North Africa not only as a food-producing and labor-using asset, but also as a main source of economic security, and social and political gain. This high amenity value, together with increasing numbers of working people in agriculture, have intensified the demand for owning or leasing agricultural land. With the exception of Sudan, the aggregate supply of cultivable land is limited and the reclamation of new lands has been monopolized by over-staffed government agencies, raising the transaction cost, reducing potential social benefits, and slowing the progress in augmenting irrigated land. Such characteristics of the demand and supply of land have tended to raise land rents, and to direct savings to bidding up land prices as a secure asset for holding wealth, particularly where the political future of big landowners is secure and when the rate of inflation is high. The preference for holding wealth in land and livestock has, in turn, tended to condition the allocation of private savings among alternative investment opportunities, both within agriculture and in the rest of the national economy.

Not only is the present supply of agricultural land limited, but its potential expansion is also limited.[3] Most lands with suitable soil fertility and moisture are presently used. A large part of the unused lands for agricultural purposes have desert soils, rocky slopes and shifting sands, with nil or very low potential for supporting rainfed agriculture. This soil texture is moisture-deficit, and it dries quickly under high temperature. Its length of plant-growing period is, therefore, very limited in one year (about 80-100 days in semi-arid areas).

In the almost rainless country, Egypt (average 0.4 inches of rain per year), the potential is only irrigable and there is no consensus on its extent. The estimates range from 0.2 million hectares (FAO, 1986) to 1.4 million ha. (Albeltagy, Egyptian Ministry of Agriculture, 1987). With regard to Morocco, the potential expansion is estimated at 0.2 million ha. under rainfed agriculture and 0.5 million ha. irrigable. If the present trend of rapid urban expansion on agricultural land, and of the high rates of entrants into agricultural labor force continues in Egypt and Morocco, the prospect for access to cultivable land is gloomy indeed. This is particularly worrying in the face of the present political environment which is not favorable to enacting further land reforms. It is projected by the above cited study, that by the year 2010 *per capita* potentially cultivable land will be one-third its 1980 level in Egypt and half in Morocco.

Sudan, on the other hand, has the largest potential cultivable area of land; 48 million ha. rainfed and 1.9 million ha. to be irrigated. By the year 2010, it is estimated that average cultivable land per person in Sudan will be 8 times that in Morocco and nearly 100 times as much in Egypt. Alas, and despite the talk about 'Arab brotherhood' and 'Arab sister countries', current international relations are not conducive to the entry of Moroccans and Egyptians into the Sudanese agricultural land market for the realization of the often proclaimed Arab 'bread-basket' in Sudan.

However, the limited supply of irrigable land requires a long time of gestation after irrigation investment. This period is between the time of the construction of the irrigation network, land levelling and the realized output growth i.e. increase in annual returns on investment over annual costs. The length of period is due partly to the prolonged duration caused by existing administrative weakness in coordination among the several government departments responsible for the diverse activities in irrigation, land reclamation and agricultural development. It is also due to the physical characteristics of soil which necessitate long period for bringing land productivity to the expected level. Sandy soil requires about 5-8 years, calcareous soil 3-5 years and clay soil 5-6 years.

Food Insecurity

North African countries are vulnerable to the threat of the food weapon by Western governments. They are not safe, and depend on the uncertain political relations and world trade regime for feeding their own people. To reduce this insecurity, they must increase food production faster than domestic demand. The task is daunting in the face of a rapid population growth and an increasing dependence on food imports. This situation contrasts with what is known from time immemorial about the region's wheat grain-production surplus, which attracted European countries to conquer North Africa.

Paradoxical as it may appear, and despite the twentieth century rapid advance in yield-increasing technology, the region has, since the 1960s, been a net importer of cereals. At present, it comprises three food-deficit countries (Egypt, Morocco and Sudan) classified by the United Nations to be eligible for food aid. Ironically, Morocco's comparative advantage in wheat production was an important factor in its occupation by France in 1912, as reported by Swearengen (1988). In the 1980s, her wheat imports represent one quarter of its total agricultural imports and food imports represent on average 35 percent of total domestic consumption. What is worrying is that Morocco's self-sufficiency rate of cereals and rice declined by 31 percent between 1969-1971 and 1983-1985 (see Table 3.2).

The data given in Table 3.2 suggest that only Libya was able, in the 1970s, to have a rate of growth *per capita* food production above one percent. The other five countries, had either a negative rates of change (Algeria, Egypt, Morocco and Tunisia), or a very low positive rate (Sudan). In the 1980s, there was a general improvement; the rates were positive above one percent in Algeria, Egypt and Morocco, but negative in Sudan and very low in Tunisia. Nevertheless, dependency on food imports is substantial and wheat imports together with cereal-aid have remained over the last two decades a permanent feature of the food situation in North Africa. In 1988, food imports as a percentage of total domestic requirements was 69 in Algeria, 47 in Egypt, 42 in Tunisia, 31 in Morocco and 12 in Sudan. Between 1980 and 1985, the percentage of aid to total cereal imports reached 41 in Sudan, 19 in Egypt and 13 percent in Morocco (*Food Aid in Figures*, 1987). Table 3.3 shows that except Libya, wheat grain and flour wheat take up, on average, 24-28 percent of total agricultural imports during the period 1985-1988.

TABLE 3.3 Wheat Imports in North Africa, 1961-1988

Years	*Wheat Grain and Flour Imports as Percentage of Total Agricultural Imports*					
Years	*Algeria*	*Egypt*	*Libya*	*Morocco*	*Sudan*	*Tunisia*
1961	13.5	14.4	23.9	15.9	13.5	35.1
1970	12.3	12.3	11.1	13.3	20.2	30.4
1975	26.5	44.9	17.0	27.8	11.6	11.0
average						
1985-1988	25.0	28.0	9.0	24.1	40.0	25.8

Source: Calculated from *"Country Tables: basic data on the agricultural sector"*, several issues, FAO, Rome

The Hydraulic Nature of Food Production Instability

The hydraulic characteristics of North African agriculture continue to influence greatly the region's food insecurity. Only in Egypt is cereal-producing land almost all irrigated (95%). In the rest of North Africa, most of the cereals (wheat, barley, millet and sorghum) are grown in rainfed areas, and the wide fluctuation in cereal output is thus closely associated with sudden year-to-year variability in rainfall. Scientists and climatologists still debate whether the recent world weather changes are signals of a new trend. Politicians, on the other hand, tend to take advantage of these changes, attributing the years of good harvest to government efforts and bad years to natural events or 'God's will'. Whereas short-term variation is likely to be caused by natural events, drought does not come as a surprise in the presence of advanced techniques on rainfall measurement and forecast.

There are other possible determinants of cereal production instability. Apart from investment rates in irrigation, they include: (1) government policy in pricing cereals far below world market prices; (2) the intervention of governments in the allocation of land and irrigation water among cereals and non- food crops; (3) the threat of desert locust; (4) political upheavals and civil wars; and (5) the extent of the intensive use of yield-increasing technical inputs, particularly the drought resistant varieties of seeds.

The cumulative effect of weather, technical and institutional determinants of instability is manifested in short- and long-term variability in the time-series analyses of yields and total harvest.[4] Based on yearly total food and cereal output data, the deviation of which from an estimated long-term trend was calculated respectively by UNCTAD (1984) and FAO (1985). The corresponding results for four North African countries are given in Table 3.4.

Table 3.4 shows an inverse relationship between the magnitude of year-to-year fluctuations and the proportion of irrigated land to total agricultural land. Although Algeria, Morocco and Tunisia have recently expanded their irrigated area, the bulk of their food production is still rainfall-dependent, and their cereal production remains highly unstable. These results contrast with Egypt's low degree of fluctuation. As noted earlier, its area cultivated by food crops, especially cereals, is almost completely irrigated.

Lastly, we need to highlight two facts established by the results of agricultural censuses and governments' sample surveys (enquête). First, with the exception of Tunisia, it is small landholders of less than 5 hectares who are the primary growers of food grain. The proportion of food crops to total area of holdings by size, is substantially higher in small holdings than larger ones (see Chapter 7). Second, there is a tendency (probably induced by governments' pricing policy) towards reducing the area under cereals, and increasing the cultivation of high-value food crops (vegetables, fruits and green fodder for livestock production). The extent of, and factors behind, the reallocation of land use among food crops in each country are examined in Chapters 4, 5 and 6.

TABLE 3.4 Estimates of Total Food and Cereal Production Instability in
Algeria, Egypt, Morocco and Tunisia, 1954-1984

Category	Algeria	Morocco	Tunisia	Egypt
1. UNCTAD - total food production coefficient of variation				
1954-1969	0.97	0.83	0.94	0.07
	(3%)	(12%)	(2%)	(95%)
1970-1982	0.43	0.93	0.30	0.05
2. FAO - cereal production	(5%)	(14%)	(5%)	(97%)
insatability index				
1969-1984	19.7	17.5	16.0	1.9

Notes: Figures between brackets are the percentage of irrigated area to total cultivated area for the last year of each period. For the definition of UNCTAD coefficient of variation and FAO instability index, see note 4 at the end of the chapter.

Sources: 1. 'Food Insecurity in developing countries: Causes, trends and policy options', a study prepared by Peter Svedberg, (UNCTAD), Table V-1, Geneva, 1984. 2. 'Food Security in the Near East: Review of the situation and measures to strengthen food security in the region', Mimeographed document no. ESPC/NE/85/3, Table 6, Near East Regional Economic and Social Policy Commission, FAO, May 1985.
Irrigated area: *Production Yearbook*, several years, FAO, Rome

Having characterized the development issues of the interlinked population growth, limited supply of cultivable land and the region's food insecurity, the discussion turns to describe briefly the institutional framework influencing these features of North African agriculture.

Governing the Rural Economy

The post-independence tendency has been an increasing role of the State in the management of the rural economy within a framework of national development planning. While we appreciate the urge for initiating the process of rural development based on the repossession of foreign owned lands and the reform of agrarian institutions inherited from colonial rule, we do not understand the continuing state control of agricultural production, marketing and trade. Invariably, there has been an extensive bureaucratization of agriculture through a wide range of government intervention, weakening the producers' incentives and motivation, and increasing transaction costs. One possible explanation is that until the post-1985 economic reforms, a combination of nationalism, Arab socialism and reformation of agrarian instititions has induced further government intervention in the agricultural sector. The second possible explanation of the centralized state authority is rooted in the form of government. The new country leadership combined legislative and executive authorities, after liberation from colonial administration (Algeria, Morocco and Tunisia), or after *coups d'état* (Egypt, 1952, Libya, 1969 and numerous *coups* in Sudan). Third, the state power in agriculture is derived from its control of the scarce irrigation water and being the largest single landowner managing state farms and deciding how much, by whom and under what terms the state-owned reclaimed land should be held. The results of the countries' agricultural census and '*enquête*' carried out in the early 1980s show that the share of state ownership, as a percentage of total arable land was 15 in Egypt, 12 in Tunisia and 7 in Morocco. In addition, the State owns forest lands and holds property rights in grazing areas. In Sudan, landed property is vested in the State.

The common feature of state control is the heavy weight of the size of state bureaucracy, backed by considerable police and military establishments. Their heavy burden on the economy is reflected in the high shares of government expenditure and services in GDP and the low shares of domestic savings.[5] However, and despite the sharp rise in government expenditure, agriculture has been neglected. There is a contradiction between what agriculture actually receives and what the policy-makers declare in national development plans, with regard to the realization of self-sufficiency in food production, and enhancing agricultural growth. In practice, agriculture has been crippled. We shall have more to say on this imbalanced development in Chapter 8.

Available incomplete information on the share of agriculture in total public expenditure and gross fixed capital formation, compared to its share in GDP and total labor force during the period 1980-1985, suggests, in broad terms, that with the exception of Libya, agriculture (including irrigation and soil conservation) did not receive the needed support by governments. Further scattered information suggests also that private investment in most of the countries has been constrained by cumbersome bureaucratic procedure. As we have noted earlier, agriculture provides employment to a major section of total labor force (between one-third and two-thirds). Likewise, it contributes from 20 to 35 percent of total GDP. Although the governments of the region committed themselves to implement the target set by the Organization of African Unity to allocate 20-25 percent of public investment to agriculture, this share in the 1980s was much lower, ranging from 7 percent in Morocco and 9 percent in Egypt to 13 percent of total in Libya and Tunisia. Most of the expenditure on agriculture is allocated to irrigation and drainage, which bring about long-term benefits.

The small private sector, on the other hand, concentrates on quick-return benefits from investment in: fertilizer, water- pumps, tractors, planting fruit trees, and production of poultry and livestock. Any expansion in investment in *large scale* irrigation and drainage schemes as well as agricultural research cannot be expected from the private sector. It should come from their governments. Furthermore, private traders and urban businessmen cannot be relied upon to undertake the needed volume and proper pattern of investment in agriculture. With regard to direct foreign investment in agriculture, the flow of foreign capital, under negotiated agreements with government, enters into profitable fields of little risk: direct involvement in exportable-cum-industrial crops but not traditional food crops; the importation and trade of farm machinery; and seed breeding and distribution.

The inadequate public investment in irrigation expansion has been compounded in the 1980s by the deepening of the countries' foreign debt and economic difficulties. This adverse effect has already been manifested in the sharp fall in the annual rates of expansion of both the total arable and irrigated lands, except in Tunisia, during the period 1980-1989 compared to the 1970s (see Table 3.2). The reality of neglecting agriculture is contrasted with the almost unaffected higher priority accorded to the non-productive sectors of government administration and military expenditure, including the purchase of arms and the armed forces' wage bill. The IMF data for the 1980s show that in the poorest country in North Africa, Sudan, military expenditure exceeds the combined public allocations for health and education.

The Bias Against Women's Participation in Agriculture

There is some considerable evidence of a common sexual prejudice against the legitimate rights of rural women in access to land, credit, technical knowledge in farming and agricultural cooperatives' membership. As household heads, farm workers, producers of livestock and dairy products, their role in agriculture, in general, and their contribution to the food sector, in particular, are underestimated. This imbalance is suggested by the results of national censuses and in programming agricultural services. It is also evident from ignoring the women's non-market economic activities (within home). Conventionally, they are excluded from valuating production and earnings in the rural economy, and in contributions to family's economic status. Rhetorical statements of ministers of agriculture at international fora are at variance with practice in rural areas.[6]

Despite an increasing awareness among policy makers of the importance of women in agricultural production and harvesting, processing and storing food products, they tend to be invisible to many agricultural extension agents, agricultural credit bankers and to land administrators. They are not reached because of legal discrimination, but mostly due to culture-based sexual bias. For example, the 1984 National Survey of Labor Force in Egypt shows that 38 percent of total working persons in agriculture were women, but their share in landholdings was only 9.2 percent according to the results of *1982 Census of Agriculture* (Table 4.6 of the published results). Another survey conducted in Sharquiyya province reveals that women's share in total labor inputs for all crops was 26 percent. However, they were responsible for most of the planting, harvesting and storage of crops, in addition to animal and poultry production (Commander, 1987, chapter 4).

Whereas redistributive land laws do not discriminate between men and woman as land recipients, the administrators interpret 'household head' to mean the male heads. They tend to assume that the accrued benefits reach all members of the household, irrespective of intra-family variation in the distribution of labor, production and processing responsibilities. In land settlement schemes, allocation of land units is almost totally confined to male household heads on the assumption that women do not possess the physical strength required in these schemes. Consequently, women's access to landed property is virtually limited to marriage and inheritance arrangements. In the privatization of communally held land, not only do women tend to lose their long-established equal rights in land-use under customary tenure, but they are also deprived of self-produced food crops. In most cases, the allotment of individualized rights in land are pro-male and pro-cash crops. As substitutes, cash- exportable crops reallocate labor to the disadvantage of women.

Most of the present agricultural extension programs in North Africa are geared towards improving the role of women as housewives and mothers, but not as farmers. Yet the female working people represent between 20 and 40 percent of total economically active population in agriculture.[7] Apart from the prevailing cultural and customary bias against rural women among agricultural technocrats, their under-enumeration usually rests on procedure. Husbands respond to enumerations on behalf of their household members, and they tend not to classify their wives or daughters as farmers, no matter how much they are involved in farming and animal husbandry.

The post-1970 rise in male migration noted earlier has far- reaching consequences on rural womens' responsibilities as temporary heads of households, both in type and amount of work. The added tasks are in addition to all year-round domestic workload, child care and livestock husbandry. Maintaining agricultural production on their absentee husbands' farming units is difficult without adequate support by local institutions (extension service, co-operative supply of inputs, lending credit and marketing the products). Micro studies on these effects and on changes in female participation in the rural labor market in affected localities are decidedly scarce. Similarly, we know little about the impact of high rate of illiteracy among female adult farmers and working girls on agricultural production and nutritional levels of children in rural North Africa. Instead, we find unhelpful generalization. The lesson learnt from past experience is that cultural constraints in rural areas cannot be altered by laws, but within a process of social change.

Rising Influence of Islamic Fundamentalists in Rural Areas

In North Africa, the State is secular, Islam is its official religion and the language of the *Qur'ān*, Arabic, is its official language. Secularism on the Western model of state institutions has, since the nineteenth century, tended to weaken the power of religious leaders and Islamic institutions. Foremost among these institutions are the very old and hitherto powerful *Ulamā* and *Muftis* (senior jurists), the Ministries of *Waqf* and Islamic Affairs, the *Shari'ah* courts and the traditional teaching center, al-Azhar (at Cairo). In some countries, mosques and the locally influential *Imams* have been brought under tight state control. In Algeria and Tunisia, Ministries of *Waqf* were abolished. The Islamic Salvation Front in Algeria was banned in 1992 and its leaders imprisoned.

Together with the influence of the Iranian experience since 1978, these developments and the spread of the Western affluent life- style have led to the rising Islamic moral force spearheaded by the revivalists' movement, particularly among the low-income groups.[8] The revivalists' influence, through local institutions in rural areas where Islamic belief is strong and the incidence of deprivation is high, cannot be under-estimated. Nor can we ignore the recently intensified conflict with governments in respect to the strict application of Islamic principles as laid down in the *Shari'ah* and practised in the dawn of Islam. In particular, the revivalists demand the prohibition of: usury *(Riba)* in interest rates payment; giving bribery to government employees; and of exploitation in land tenure arrangements,trade dealings and other economic transactions (see Chapter 2, sections 4 and 5).

Commonly known in the Western media as fundamentalists, the extremists question the legitimacy of the worldly Westernized State for administering Moslem communities and, in turn, the body of legislation issued by state institutions. They claim that such a secular structure does not represent the Divine Sovereign revelation and that this structure was created during and after long rule by European countries whose values and style of life are alien to Moslem values, particularly mortgaging land, wine-making and drinking alcohol.

Governments have differed in their response to the movement's demand. Some have authorized the establishment of Islamic Banks, and amended existing laws on property rights, credit and banking systems to conform with these principles. In Sudan, the fundamentalists' leaders became senior members of the government since 1981, and have orientated the legal framework, abolishing payment of interest rates in all transactions of the Agricultural Credit Bank. Others have banned the activities of this movement and imprisoned its leaders on the grounds that such movement threatens the country's social unity and security. Nevertheless, their popular support in rural areas is gaining strength.

Notes

1. Independence years are: Egypt 1923; Libya (Libyan Arab Jamahiriya) 1951; Morocco 1955; Sudan 1956; Tunisia 1956 and Algeria 1962.

2. Estimates for Egypt and Libya are calculated by the writer from areas expropriated and redistributed. Those for Algeria, Morocco, and Sudan are based on data given in El-Ghonemy, ed., 1967 and several country progress reports.

3. This statement and the estimates on potential expansion are based on soil studies on potential cultivable lands until the year 2010 carried out by the Food and Agriculture Organization of the United Nations (FAO). They are summarized and presented in *"African Agriculture: the next 25 years"*, Tables 3.02 and 3.16, *Atlas of African Agriculture*, 1986.

4. The degree of instability is relative to an assumed state of stable or normal growth. Measurement of instability is sensitive to both the accuracy of data and the index used. The UNCTAD's study used the sum of the square of deviations of actual food production in a given year from the trend value, divided by the sum of the square of the deviations from the average production of the number of years (n) adjusted for degrees of freedom n/n-2, the result being the coefficient of variation. The index of instability used by FAO is 100 multiplied by the deviation and divided by the actual production in a given year. The exact measurement of instability in food production is useful to policy makers in the formulation of food security policy and the assessment of required level of the stock of food reserve, and of imports.

5. See *'development indicators'* in *World Development Report* 1988 and 1989, and Volumes VII-IX of *Government Finance Statistics Yearbook* (IMF). The share of government services in agriculture and the cost bill of government employees in the agricultural sector are not disaggregated.

6. These government statements on the importance and equal rights of women in agriculture were given at the eighteenth Regional Conference for the Near East (including North African countries). The conference which was held in Istanbul in March 1986 was attended by Ministers of Agriculture. The item discussed was "The role of Women in Food and Agriculture in the Region".

7. Most of censuses and labor surveys underestimate rural women's time allocated for productive work in agriculture which is used as a criterion in the classification of economically active population. For example, in Egypt, the population census of 1986 estimated the female share in the labor force in agriculture at 19 percent while the labor survey of 1984 shows the share as high as 38 percent.

8. On the Western concern over fundamentalism, see Wright, 1992.

4

Transformation of the Rural Economy: Algeria and Tunisia

From the preceding chapter, the reader now has an idea about the broad structural characteristics of the Maghreb countries (Algeria, Morocco, Tunisia and Libya). Each country's development indicators, given earlier in Tables 3.1 and 3.2, are influenced by a complex interaction of historical, economic and institutional factors, as well as natural events.[1] Foremost among the institutions, influencing the producers' incentives and distribution of income in the rural economy, are state authority, landed property rights and land-based power structure. They were left to empirical examination in an historical perspective. This is the task of the present and next chapters. The emphasis is on the sequential policy-making, which has shaped the present structure of production and the distribution of wealth and income in agriculture.

In order to maintain a balance of presentation among the chapters, Algeria and Tunisia are the subject of this chapter, leaving Morocco and Libya to the next. The proposition underlying this subdivision is simply that the duration of colonial rule has shaped the rural economy's agrarian institutions in different degrees, which significantly influenced the content of rural development strategy in general, and land policy, in particular, after independence.

Algeria

The process of colonizing land and subjugating state institutions in Algeria was started by France much earlier than the other Maghreb countries in the last century. Not only was the colonization of Algeria the earliest, but the country's entire economy and administration was made part of metropolitan France.

Integrating Agriculture into the French Economy, 1840-1960

Since the flow of French emigrants in the 1830s and the 1840s, the policy of France aimed to virtually integrate the Algerian economy and landed property legislative framework into those of France. According to laws of 1844, 1846, 1851 and 1873, communally held tribal lands were partly sequestrated for colonization purpose, and partly registered in individual titles granted to Moslem families who collaborated with the colonial regime. The Islamic institution of holding lands under trust *(waqf* or *habous)* was changed, and its lands were purchased in the 1830s and 1840s, and put under the disposal of French settlers *'colons'*. In this way, *waqf* was practically abolished and the old-established communal lands of the tribes were restricted. In addition, all forest and uncultivated lands were expropriated and converted into state ownership, and lands owned by the Turks were confiscated. Thus state domains became substantial, amounting to 3.1 million hectares (Issawi, 1982, pp. 39-40).

Land Concentration

By 1954, the French settlers numbering about 23,000 owned 2.6 million hectares largely in the fertile land. Situated in the North *(Tell)* with high rainfall, the French-owned land amounted to nearly 30 percent of the total cultivable area. As Table 4.1 shows, 87.4 percent of the area of non-Moslems' holdings was in this top category, with an average size of 373 hectares per landholder. Some European land companies owned more than 10,000 ha. each. Such a concentration of wealth contrasted sharply with the overall average size (11.6 hectares) for all Moslem holdings, 95 percent of which were below 50 ha. in size. They were mostly situated in the less fertile slopes of mountains and the low rainfed areas.

Output value of the French-held lands which grew vineyards, citrus, sugar beetroot, tobacco, vegetables and cereals was nearly 7-10 times that of Moslem-held lands. Hence, a wide disparity in yields and incomes was created between the fertile lands of the French *colons* and the rest. Although the *colons* represented about 3 percent of total population, their income share was disproportionately very high; 60 percent of total gross agricultural income and 52 percent of net profit from agriculture (Griffin, 1976, pp. 26-28). The Arab farming population provided cheap labor at subsistence wage for the commercialized and labor-intensive French farms.

TABLE 4.1   Pre-Independence Distribution of Land Holdings by Size and
             Ethnic Category in Algeria, 1960

Categories of Size (hectares)	Number of Holdings			Area of Holdings (thousand hectares)		
	Moslems	Non-Moslems	Total	Moslems	Non-Moslems	Total
Less than 10	438,483	7,432	445,915	1,378.4	22.6	1,401.0
10-50	167,170	5,585	172,755	3,185.8	135.3	3,321.1
50-100	16,580	2,635	19,215	1,096.1	186.9	1,283.0
100 and above	8,499	6,385	14,884	1,688.8	2,381.9	4,070.7
Total	630,732	22,037	652,769	7,349.1	2,726.7	10,075.8

Source: *Statistique Générale de l'Algérie, Tableaux de l'Économie Algérienne,* 1960, p.
129, Cited in Sayigh, 1978, p. 525. Reprinted by permission from the Publisher, Groom
Helm (now Routledge).

Transforming the Rural Economy on Socialist Lines

After gaining independence in July 1962, the Algerian leadership focused on
the nationalization of the farms formerly owned and managed by the French
settlers. Production in this economically important sector suffered from the war
of liberation. Also, it laid the foundation for a socialist economy.

The Formation of the Socialist Sector

The recovery of a total area of 2.6 million hectares was spontaneously
achieved by the sudden departure of foreign settlers. Supported by the country
leadership, the Algerian workers in these farms took over their management.
Thus, production was resumed, and the break-up of large farms avoided. This
nationalized sector was called *exploitation d'auto-gestion agricole* or
self-managed farms and, in the 1970s, became known as the socialist sector.
Some 2,500 committees of workers were organized, each comprized 7 members,
on average, to manage the farms, within the framework of the national
development plan. The area of each farm ranged from 100 ha. of vineyards,
citrus, sugarbeet (beetroot) and vegetables to 2,000 ha. of cereals. Farming was
undertaken, as in state farms, by salaried workers, while the State retained landed
property. The net profit was partly invested at the farm level, and partly
deposited in the 'National Investment Fund' established to finance the country's
economic development program. By 1966, nearly one million working persons
were absorbed in this large socialist sector (El-Ghonemy, 1968, pp. 40-1).

In 1964, the landed property of Algerians *(bachāghā)*, who collaborated with the French regime prior to independence was confiscated. It amounted to nearly 100,000 hectares of cultivated land. A network of agrarian reform co-operatives were established and their membership was obligatory. Together with the self-managed farms, these co-operatives were linked to the central government's Agrarian Reform Office (ONRA), which arranged the marketing of products in both the domestic and foreign markets (Raggām, 1967).

Restructuring the Traditional Sector

So far, the traditional sector of nearly 6 million hectares (including 4 million ha. cultivable land) was left out of the socialization program. So were the Algerian Moslem large landowners and the extensive area of state-owned land. This large sector was rainfed and produced most of the cereals and animal products. It comprized large and small landowners, including the *fellaheen* who were poor tenants, share croppers and landless wage workers. They were estimated at 920 thousand comprizing about half a million landless workers (Pfeifer, 1985, p. 48). They had high expectations awaiting realization by the repeatedly spoken of *"La réforme révolutionaire"*. The promise was operational by President Boumediene's Charter of Agrarian Revolution issued in November 1971, and implemented between 1972 and 1976. The Charter[2] provided for the redistribution of 0.8 million ha. of state-owned land to tenants and to a section of the landless agricultural workers.

It provided also for the expropriation, without compensation payment, of all private landownerships belonging to absentee owners who depended for living on other sources of earnings. This act affected units exceeding the established limit of value productively -- approximately 40-50 hectares on average. After some exemptions were made by local politicians, nearly one million hectares were expropriated, including 36,000 ha. irrigated. The total area was redistributed in units of 10-15 hectares. Land recipients had the right of use but not ownership which was vested in the State. The criterion employed for determining the unit size was that it should provide 3,500 dinars in 1975 (one dinar equalled US$ 5). The rule was that two-thirds of the beneficiaries should be small tenants (holding less than 5 hectares) and one-third landless wage workers. They were automatically grouped by the government into collective production co-operatives (CAPRA) that represented 72 percent of total (see Appendix Table 3).

The revolutionary reform was extended in 1975 to benefit the pastoral nomads and the semi-nomadic people (agro-pastoral). Those who were living in a semi-arid fertile areas and communally held land were settled in individually marked holdings and were also grouped into production co-operatives. Weexteen (1977, p. 195) reported that all tribal beneficiaries continued to be responsible for livestock production, palm-groves and cereal cultivation in low rainfed lands (100-300mm rainfall per year). This phase of the agrarian reform was named

"La Révolution Pastoral''. Its aim was to improve the earnings of agro-pastoral people through the provision of livestock health services, and the modernization of production and marketing arrangements in pasture.

The several categories of beneficiaries of this comprehensive reformation were required to join co-operatives, having different names and scope of functions, but the agrarian revolution co-operatives (CAPRA) were dominant. These government-controlled institutions channelled the subsidized inputs and procured the marketable surplus for sale and export through the state organs. Whereas the participation of the beneficiaries was administered, the voluntary services of rural youth among university students and the National Algerian Peasants Union (UNPA) effectively contributed to the implementation of this program.

Impact on Food Production

Our account of the Algerian strategy for rural development during the 1962-1982 period should be seen in the context of an overall centrally planned economy organized along socialist lines. Despite the state control of investment activities, the institutional transformation of agriculture had not been backed by adequate investment to meet the production requirements. This deficiency occurred during a period which witnessed a rapid flow of revenues from oil and natural gas exports, especially after the 1973 and 1979 sharp rise in world prices. Also, it manifested itself in spite of the sizeable remittance receipts from Algerian migrants to France (US$ 447 million in 1982).

The consequences were serious. Irrigation expansion was very slow. With expanding irrigated land between 1962 and 1980 at the meagre rate of 0.5 percent *per annum*, the degree of instability of cereal production was high at 20 percent and wheat imports reached 27 percent of total imports in 1975 (see Tables 3.3 and 3.4). Cereal production instability was clearly manifested during the droughts of 1966/67 and 1973/74. The deterioration in food production was reflected in falling real income per working person in agriculture, together with a decline in total and per head food production (see Table 3.2 for the period 1970-1980). Likewise, the area of wheat, the principal food, diminished from an average 2,017,000 hectares in 1965-1969 to 1,770,000 ha. in 1979-1982. Throughout the entire period (1965-1982), average yield of wheat remained low around 0.6 ton per hectare, which is below that in Algeria's neighbouring countries during the same period. The yield was far below the estimated potential at 2.0-2.5 ton/ha. (FAO Data Base). Thus the politically dominant socialization of agriculture brought about food insecurity.

Post-1982 Reforms:
From Socialist to Capitalist Agriculture

Having attained an egalitarian rural economy since independence, the policy-makers realized that it could not be sustained by low productivity and inadequate producers' incentives. Subsequently, in the Ninth Development Plan (1980-1984), followed by the Tenth Plan (1985-1989), the government directed oil-revenue for rural development: it increased the share of agriculture in total investment from 11 percent to 14.4 percent, and devoted half this allocation to irrigation. Similarly, public expenditure on health, education and the construction of new villages was raised by 30 percent.[3] Agricultural and rural development bank (BADR) was established, and linked to *Albaraka* Islamic Bank for lending, according to Islamic principles. The management of agrarian institutions was decentralized. The aim was to decontrol the rural economy, and to enhance the productive capacity of agriculture and the quality of life in rural areas in order to check the rapid rural-urban migration.

TABLE 4.2 Index of Changes in Irrigated Area, Food Production and Import Dependency Ratio in Algeria, 1960 - 1989

	1960	1970	1980	1988-1989
Irrigated area	100	106	113	159
Harvested area of cereals	100	81	77	80
Cereal production	100	219	258	211
Meat production	100	143	237	277
Total food production[a]	68	85	107	120
Food productivity per person[b]	115	116	107	90
Volume of food imports[c]	33	23	109	189
Food imports dependency ratio: average 1969-1971 = 32.1%				
average 1986-1988 = 70.7%				

Notes: a, b, c are based on FAO index: average 1979-1981 = 100. Cereal, meat, and total food production indexes are based on volume of production in metric tons. See definition of imports dependency ratio in Table 3.2.

Sources: Calculated from data given in *Production Yearbook*, volumes 17-43 and *Country Tables: basic data on the agricultural sector*, 1989 and 1990, FAO, Rome. Import Dependency Ratio is from *Human Development Report, 1991*, Table 13 of indicators, UNDP, Oxford University Press.

Faced with the *fellaheen's* rising discontent about the tight state control of production organization and the inefficient management of co-operatives, the new privatization and liberalization policy has gradually converted land use rights into private ownership. The post-1982 policy has also increased the producers' responsibilities in production, marketing, savings and in private investment. These new policy shifts were reinforced by Law no. 9 of December 1987 for the individualization of collective production co-operatives, and for the privatization of land and means of production. Moreover, about 3,000 state farms were sold, in private units, to individual farmers at favorable terms. The purpose has been to enhance the *fellaheen's* production incentives, improve the economic efficiency of family farm units, and to secure the *fellaheen's* command over food intake.

Some progress has been achieved since the introduction of these policy changes. Table 4.2 shows a significant increase in irrigated area and meat production. The rise in animal products is reflected in the increase in food production index by 13 percent between 1980 and 1989. However, the production of cereals has been poor. Its index in 1989 fell below its 1970 level, and with continuing high population growth at an annual rate of 3 percent, food production per head fell. Hence, the volume of food imports sharply rose by 80 percent in the 1980s. Domestic food production remains a major problem.

Considering that agriculture is still the largest sector in the Algerian economy with respect to employment, the indicators of agricultural performance suggest that its production techniques have not been sufficiently transformed in order to match the fundamental changes introduced in land tenure system. Although the share of agricultural output in total GDP and in foreign trade has been dwarfed by oil and natural gas, nearly one-fourth of total population are employed in agriculture. Their productive capacity is still low, despite the necessary but insufficient irrigation expansion. Wheat, the staple food, occupies nearly half the cropped area, but its average yield being at 650 kilogram per hectare is not only far below the potential noted earlier, but also the average of 1.1 ton/ha. in Sub-Sarahan Africa. The yield has increased by only 10 percent over the last two decades. Likewise, the rural-urban income gap is still wide, resulting in the rapid urban population increase at the average annual rate of 5 percent in the 1980s, compared to 3.5 percent in the 1960s.

What policy changes are likely to be introduced following the Islamic fundamentalists' *(Front Islamique du Salut)* growing public support[4] in 1990-1991 checked by military action in 1992, remain to be seen.

Tunisia

The question of holding agricultural land and its political and social implications have been central to the historical experience of Tunisia, since the Arabs ousted the Romans in the seventh century. In consequence, the principles of inheritance, land tax *(ushr* and *Kharāj)* and holding landed property under the

Islamic *waqf (habous)* were introduced. Subsequently during the long Ottoman rule (1569-1881), land taxation was excessively increased around the middle of the last century which triggered rural unrest and the *fellaheen's* revolt in 1864 (Perkins, 1989, p. xi and p. 8). The economic situation was worsened by devastated cereal crops and the periodic outbreak of cholera and famine in rural areas in the 1860s. This human disaster forced many *fellaheen* to flee to towns, and pressed upon the rulers to borrow heavily from French financing institutions (Issawi, 1982, pp. 97-100). The coalition between lending institutions of France and other European creditors influenced the French government to occupy Tunisia, which was declared a Protectorate in 1881.

The Process of Colonizing Rural Tunisia, 1881-1956

The rural economy of Tunisia was shaped late last century by the colonization of the richest lands to grow wheat and grapes for making wine. The process was accelerated in the early part of the present century, resulting in the dislocation of many native farming people who and their flocks were pushed to the less fertile land and to the South, where average rainfall did not exceed 5-6 inches *per annum*.

Duality, Inequality and Poverty

By 1956, when Tunisia gained independence, some 800 thousand hectares of fertile land (20 percent of total arable land) were held by 6,500 Europeans, mostly French. Out of this total area, 600 thousand ha. was owned, and the rest was held under rental arrangements. These actions formed a modern sector, in which the average size of farms ranged from 100 to 200 hectares per settler. They were predominantly located in high quality rainfed lands where the mechanized wheat farms, modern irrigation methods and high-value vegetable crops and fruits were introduced.

Reliable data on living conditions in rural areas, at the time of independence, are scarce. The census of 1956 tells us that the population was overwhelmingly rural (70%), and 57 percent of total workforce depended on agriculture for employment, out of whom 25 percent were classified as 'unemployed'. Chebil (1967) reported that although there were a few modern Tunisian farms whose owners were rich, most of the Moslem Tunisian working force in agriculture (0.7 million) were poor. The rural poor comprized small landholders in rainfed areas, sharecropping peasants, landless workers and nomadic-pastoralists. The latter were engaged in livestock raising and the cultivation of cereal crops, when rainfall permitted. Nearly half the working force in agriculture were landless laborers.

About one million hectares, not alienable in the land market, were operated under the Islamic land tenure arrangements of *waqf (habous)*. Ironically, and contrary to Islamic rules, these lands were poorly managed, starved of capital investment, and tilled mostly by insecure tenants and landless squatters. Consequently, its cultivators' disincentive to invest in land improvement prevailed and, in turn, its cereal production had deteriorated. Approximately, 2.5 million hectares of marginal lands in the central and Southern regions were held customarily by semi-nomadic tribes under communal ownership. Chebil also reported that poverty prevailed in the 1950s among the *fellaheen* in the *Habous* lands, and in areas of the rented out lands by Moslem Tunisians.

These agrarian conditions were manifested in low annual income per working person in agriculture estimated in 1966 by the World Bank at 84 Tunisian dinars (about US$ 210), or less than one-fifth of that of non-agricultural labor force. Several estimates were made about the incidence of rural poverty in 1966; each used different definition. They range from 20 percent by the World Bank, to 49 percent by ILO, and even higher at 60 percent according to van Gunneken's estimation (Radwan *et al* 1990, Table 4.10).

Restructuring the Agrarian System for Rural Development

As might be expected, the reform of the appalling conditions of rural underdevelopment started immediately after independence in 1956. The pressure for rural reforms did not come from the *fellaheen* who were neither organized nor had lobbying power. It came from the middle class activists in the *Destour* Party which led the country to independence (Simons, 1970, p. 17). There was a pressure from the labor union's leader Ahmad Ben Saleh for instituting a radical reform along socialist lines (Perkins, 1989, p. 135). Yet, the emphasis in the deliberations of the *Destour* Party Congress at Sfaqis in 1955 and the Ten-Year Plan for development were placed on the modernization and diversification of agricultural production, and the expansion of employment opportunities in rural areas. It seems that there was neither an urge among the policy-makers to fix a maximum ceiling on private property in land, nor to antagonize the rich landed élite. Likewise, there was no tendency to antagonize foreign landowners, and to regulate the land-lease market.

Reforming Land Tenure

Unlike Algeria, institutional changes in the Tunisian agriculture began with the abolition of the archaic institution of public *waqf or habous* by Laws of 1956 and 1957. The State acquired their property and abolished the Ministry of *Waqf*. Private *habous* lands were divided into individual ownership units among the legitimate claimants. Clearly, by this course of action the State rejected the Islamic *waqf* tenure, as not being conducive to rural development. This reform affected 29 percent of total arable land of 4.2 million ha., excluding the areas of forest and natural pasture. This extensive area was gradually transformed into productive farming enterprises, through irrigation investment and improvement of cereal and livestock production.

Next came the regulation of the communally held tribal land rights by Laws of 1957 and 1959. Most of these hitherto underutilized lands were changed to individual ownership of settled agriculture in units of 10-20 hectares each, mostly with olive plantation, and the rest remained as communal ownership. In both cases, farmers were organized in co-operatives providing them with technical assistance, construction of tube-wells and basic means of production to diversify and intensify land use according to rainfall. Government-patronized production and service co-operatives were organized by Law of May 1963. Grazing lands were also assigned to co-operatives to control the numbers of livestock and to improve pastures. Certificates of possession were granted by the State to individual landholders giving them the right to borrow from the State-owned agricultural credit bank. Consequently, the production of cereals as well as livestock and olive products became market-orientated.

Limiting the maximum size of landownership at 50 ha. was confined to the newly irrigated areas of Mdjerdah Valley scheme by the Law of June 1958. The area exceeding that limit was expropriated against payment of compensation in the form of guaranteed Treasury bonds. It was distributed in ownership units of 5-10 hectares to small tenants having experience in irrigated farming. The criterion for unit size was to provide a *minimum* income of 250 dinars per family per year (US $ 600 in 1965) from growing vegetables, fruits, wheat and from dairying (Parsons, 1965). The new owners were required to pay for the land over a period of 20 years and to join production co-operatives which became the sole source of credit and the agent for marketing the products. One notable characteristic of this policy is the requirement that landowners' share in the costs of publicly funded irrigation works, according to land productivity and its value. The charge ranged from 25 percent to 60 percent of value added (Chebil, 1967, p. 196).

An important instituted agrarian change was the scheme for the consolidation of scattered small parcels of land. This scheme and the above acts of reformation have laid the foundation for technological change and rural development. Apart from the mandatory membership of co-operatives, the government provided rural population with primary education for boys and girls, health and child-care services, housing, adult literacy, and skills formation for the youth. A special program was initiated for rural women which provided them with training in housekeeping, handicrafts and nutritional education.

Experimenting with Collective Production Co-operatives

The recovery of the already developed and commercialized foreign-owned farms took place 8 years after independence. By force of the Agricultural Land

Property Law of 1964, nearly 16 percent of total cultivable lands or 700 thousand hectares of the best lands in the country were taken over by the government. The acquired land was partly purchased at the market price, and partly nationalized. A small portion of it was allotted to landless farmers and the rest was retained by the government and managed as state farms.

Throughout the countryside, farmers holding individual ownerships *(melk)* and leased lands were grouped into collective production co-operatives which were virtually managed by a hierachy of bureaucracy. Experience has shown that the hurridly implemented mandatory co-operativization did not work, particularly where individual land titles were granted. They were short-lived, and in 1969 were dissolved after a widespread and growing discontent among farmers with the unsatisfactory results of this imposed system. Since 1969, voluntary membership has replaced the compulsory adherence and individual management responsibility of private farms has prevailed.

The Impact on Food Production

The above account of the institutional transformation during the period 1956-1970, combined with public efforts to modernize production and intensify land use have had their effects on productivity and the distribution of income in rural Tunisia. Through this process, the reliance of agriculture on both public investment and the government's administrative capability was strong. Whereas gross inequality in land tenure was reduced, the production performance was poor, particularly in the 1960s.

From the author's discussions with co-operatives' leaders and officials of the Ministry of Agriculture during his visit in 1976, it appeared that two factors were behind the stagnation of agriculture in the 1960s. One was the government's experimentation with collective production co-operatives that was resisted by farmers, whose resentment, disincentive to work hard and resignation prevailed. The other was the droughts of 1961-1962 and 1967, reducing wheat and barley harvests to less than half their levels in the preceding years.[5]

The 1960s' sharp fluctuation in the production of wheat and barley, the two major crops, is shown below in thousand metric tons:

	1960	average 1961-1962	average 1964-1966	average 1967-1968	1970
Wheat	524	213	716	360	519
Barley	236	50	288	100	151

Average yield of wheat fluctuated also between 500 and 800 kilograms per hectare during the 1960s, which affected the producers' incomes and nutritional standards. Wheat, both hard *(durum)* used for bread, and soft used for making *couscous*, is the principal staple food, providing nearly 60 percent of total calorie-intake. Alarmingly, food productivity per head fell in the 1960s at the average annual rate of a negative 2 percent (FAO *Country Tables*, 1989, p. 276). This unsatisfactory food production was manifested in low annual rate of agricultural GDP at 2.0 percent.

This poor production performance occurred, despite an expansion in irrigated land in the 1960s by 25,000 ha. The records of the Ministries of Planning and Agriculture showed that the share of agriculture in total investment was between 19 and 22 percent, and that half of this allocation went to irrigation and soil conservation. However, the low output growth in the 1960s is not surprising because of the long period of gestation required for bringing about a rise in yields and total output. It was also learnt from the Ministry of Agriculture that high-yielding varieties of wheat (HYV) combined with improved producer's price were introduced only around 1970. Subsequently, the production performance has improved.

During his visits to Tunisia in 1984 and 1985, the author examined post-1970 changes in the productive capacity of agriculture. Between 1970 and 1985, substantial progress was made towards rapid irrigation expansion. Likewise, the government introduced, and subsidized high-yielding varieties of wheat, and stimulated the private sector's investment in agriculture. In consequence, irrigated lands expanded by 160 percent between 1970 and 1985, yields of wheat rose by 30 percent on average, and agricultural GDP annual rate of growth doubled to 4.2 percent in 1970-1981 compared to its rate in 1960-1970 (*World Development Report*, 1983). Who benefited, and how much, from this pattern of agricultural growth depends chiefly on the state of access to land and its size distribution, resulting from the land policy instituted in the 1960s.

Distributive Consequences

The 1980/81 Survey of landholdings conducted by the Ministry of Agriculture shows that 85 percent of total farming population had access to land (owners, tenants, share-croppers, and agro-pastoralists). Nearly one-fifth of agricultural families were left out by the several agrarian reforms as landless wage workers. Since then, some have migrated to work in Libya, and some have been absorbed into the government-promoted non-agricultural activities within rural areas as well as in the expanding urban centres of the economy.

Table 4.3 shows that out of total arable land, the major part (87.6%) is privately owned *(melk)* in units of 13 hectares on average. The state-owned land representing 12.4 percent of total includes 48 state farms; the average size of which was 4,500 hectares. Furthermore, the findings of the agricultural survey indicate a high inequality in the size distribution of *melk* lands.

TABLE 4.3 Ownership of Arable Land in Tunisia, 1980/81

Land Tenure Status	Area in ha	%	Number of farms
Private-owned land *(Melk)*	4,452,369[a]	87.6	350,000
State-owned land of which:	632,369	12.4	5,000
	5,084,738	100.0	355,000
Agro-combinants, pilot farms and land operated by the State[b]	216,007		48
Production co-operatives[c]	236,664		228
Land assigned to agricultural research and regional institutions	56,254		26
Land rented[d]	61,446		4,698
Grazing land where collective access and use are practised[e]	61,998		

Notes: a Not including grazing area owned and operated collectively. b The "Office des terres domaniales" (O.T.D.) is the state agency responsible for managing this land located mainly in the Northern part of Tunisia. c Located in Nabeul, Beja, Jendouba, Siliana and Bizerte provinces. d These are scattered small holdings rented out by the State before their eventual sale. e Mainly in the Central South.
Source: Based on the results of the Agricultural Survey *(enquête)*, 1980/81, Ministry of Agriculture, Tunis. Original in French and Arabic.

Out of the total area of privately owned land, 7.4 percent is in the size group of less than 5 hectares, the owners of which represent 43 percent of the total number of owners, and their farms' average size is 2.2 ha. According to the results of the survey, this large section of farmers used only 12 percent of total consumption of chemical fertilizers, while large farmers in the size category of 50 hectares and more used 55 percent of total *(Enquête Agricole de Base, 1980/81)*. In the top category of 100 hectares and over, about one percent of private owners possess 17.5 percent of total land *excluding* state-owned.

In addition, the *'enquête'* reveals that it was not wheat which benefited from irrigation investment (only one percent). Rather, lands growing high value crops such as vegetables, fruits and forage for livestock production did. The latter are mostly cultivated by large farmers (50 ha. and more). Thus the benefits from the introduced technological change and government subsidization policy in agriculture accrued mostly to large farmers.[6] The National Institute of Statistics (INS), analyzed the results of the 1981 agricultural survey and reported that whereas small farmers in the size category of less than 10 ha. represented 63 percent of total number and 19 percent of total area, they received 13 percent of the subsidized HYV seeds of hard wheat. On the other end of the scale, large farmers holding 50 ha. and more each obtained 43 percent.

Thanks to the spread of non-farm activities in rural areas and to the remittances received from rural migrants that poverty levels were reduced (see Chapter 7). However, the INS' analysis indicates that the incidence of absolute poverty and undernutrition in rural areas is higher among the landless wage workers than other occupational groups in agriculture. The landless workers represented 37 percent of total rural poor and 77 percent of total adult illiterate in rural areas.[7] No wonder that the government launched in 1988 the national campaign against poverty *"Lutte contre la pauvreté"*, the results of which would be judged in the late 1990s.

Notes

1. For a detailed review of these interacting factors, see 'Land Tenure' in Chapter 'Agricultural Expansion', Issawi (1982).

2. These elements of the program are obtained from the Ministry of Agriculture and Agrarian Reform's published set of legislations, in Arabic.

3. *"Deuxième Plan Quinquennal, 1985-1989, Rapport Général"*. Algerian Ministry of Planning and Development, January 1985.

4. This Algerian experience represents a case test for reconciling Islamic fundamentalism with genuine democracy. Before the introduction of multi-party system in 1990, several leaders of the Islamic groups were killed, in 1987, by the security forces. Through a free election, the Islamic Front (FIS) won 61% of local councils in 1990 and 82% of the parliamentary seats in December 1991. By military actions, taken in the early 1992, the FIS leader Dr Abassi Medani, a university professor of philosophy, and his deputy were imprisoned, the elected local councils dissolved, and FIS banned.

5. *Production Yearbook,* vol. 21 and 24, and *1989 Country Tables,* FAO, Rome.

6. For a detailed discussion on the effects of pricing policy on landholders' economic gains by size, see Radwan *et al,* 1991, pp. 36-44 and the World Bank Study by Cleaver, 1982, pp. 45-53.

7. Socio-economic indicators, 1982, INS, Tunis, p. 58, in Arabic.

5

Contrasting Rural Development Strategy: Morocco and Libya

The preceding chapter began with an explanation of why the duration of colonial rule is used as the criterion for dividing the four Maghreb countries. Chronologically, Algeria and Tunisia were colonized by France during the nineteenth century, while Morocco and Libya were respectively subjugated to the French and Italian rule in the early part of the present century. This chapter examines land policy, food production and rural development experience of Morocco and Libya. Since independence, each has followed a distinct path towards the realization of development objectives with respect to greater equity and food security. As we shall soon find out, the historical factor, shaping pre-independence agrarian structure, is the common element behind the design of their respective rural development strategy.

Morocco

The story of Moroccan experience differs from that of Algeria and Tunisia in origin and the path followed. The duality of the agrarian structure, associated with the formation of large private estates, started long before the French occupation in 1912. Tracing this origin helps to explain the current concentration of wealth in rural Morocco.

The Origin of Large Estates' Formation

With their increasing profits from trade between Morocco and Europe, the rich merchants of Fez expanded their wealth in rural areas through holding large tracts of agricultural land. It seems that the motive behind their land accumulation was to gain political and economic power. The power was in the control of cereal supply and trade in rural Morocco in order to influence the central government *(Makhzan)*. They succeeded, and their powers were reinforced by receiving land grants from the Sovereign *(Moulay)* who held the absolute property rights of state-owned lands.[1] Large areas of 300 to 600 hectares were granted to influential families from the cities of Fez and Meknes, senior government officials, members of the *Moulay* family and to Moslem leaders *(Shorafā, Ulamaā and Qādi)*. Through legal manipulation, the usufruct rights in the granted lands *(Iqtā and Azib)* were gradually converted into private property *(Melk)*. Lazarev (1977) gave a detailed account of this process of property transfer, and listed the names of recipient families, many of whom are presently dominant in Moroccan agriculture and in the national political arena.

In addition, the influence of these families on the bureaucracy enabled them to extort additional areas of cultivable land belonging to small landholders *(paysannerie)* and to register the grabbed lands *(ghasb)* as their private properties. They also took advantage of the sale of state-owned land on favorable terms, following the government's financial crisis caused by the high public spending on the Spanish-Moroccan War of 1860. Issawi (1982, p. 97) reported that the economic crisis was deepened by frequent droughts and the outbreak of plague and famine in 1858 and during the 1878-1882 period. Numerous hungry and dispossessed small peasants fled their land, seeking food security in towns. They sold their land and livestock at very low prices (Lasarev, 1977, pp. 84-85). The rest of the land, partly growing wheat and barley, and partly used for grazing, was held collectively by tribes including the two powerful tribes: the *guich* in the Western planes and the *berbers* in the middle Atlas mountains.

The Colonial Agricultural Policy, 1912-1955

Against this background, the French Protectorate began in 1912. Before that year, the Moroccan agricultural development potentials were carefully studied by several groups of French politicians and technicians. Their systematic work laid the foundation for the colonial agricultural policy and for the estimation of the expected benefits to France from the distinct character of Moroccan agriculture. These characteristics comprized the extensive fertile planes in the West and North near the Atlantic and the Mediterranean coasts, the rich water resources for potential irrigation work between the Atlas mountains and the Atlantic coast, and the availability of much cheap labor of landless-wage workers and sharecroppers. The potential wheat and citrus output growth was instrumental in the formulation of the colonial policy which has, since then, shaped the structure of agricultural production and the rural system. Its main features should, therefore, be understood.

The historical record suggests the following policy features which are relevant to our study.

1. Harnessing water resources by way of constructing dams and canals for irrigated wheat, citrus and high-value vegetables to be regularly supplied to France.
2. The appropriation of landed property for both official and private colonization, not to be on a large scale as followed in Algeria, but at a relatively limited scale.
3. The sale of land acquired for *official* colonization to selected individual and corporate French settlers in large holdings of 200-300 hectares.
4. Bilateral arrangements in land transactions for *private* colonization purpose between French and other European settlers, on the one hand and the Moroccans, on the other.

The Pre-Independence Agrarian Setting

By 1953, there were 4,270 *private* settlers owning 728,000 hectares with an average 200 ha. each. The colonized land was mostly situated in Casablanca and Rabat regions of the Chaouia and Gharb planes near the Atlantic Coast as well as in the Mediterranean plains of Basse Moulouya. In all, some 6,000 French settlers and foreign companies (both official and private) owned nearly one million hectares of fertile and irrigated lands, most of them around the cities of Fez, Meknes, Rabat and Marrakesh (Swearingen, 1988, Table 8). More than half the total perennially irrigated land was owned by European settlers who represented only 5 percent of total landholders. They dispossessed Moroccan Moslems, despite a strong opposition from the Sovereign *Khalifa Mawlai Abdul Hafed* and the Moslem leaders against the foreigners to own agricultural land.[2] As remarked by a French scholar who lived in Morocco before independence "the people could only see the transfer of their land to foreigners, nicely disguized in formal judicial phrases" (Pascon, 1986, p. 86). The majority of Moslem farmers numbering about 900 thousand families lived from 6 million rainfed hectares, mostly for grazing, livestock-raising and rainfed cultivation *(bour)*, which grew cereals under the fallow rotation system. Half the total number of their holdings was less than 2 hectares in size. Landless workers were about one-fourth of total agricultural households, and another 15 percent possessed less than 0.5 hectares each, many of whom were also hired laborers. There were also rich Moslem farmers owning more than 50 ha. each, and many of them benefited from irrigation investment during the French Administration.

As noted earlier, they were already dominating the rural economy before the French rule.

Food Insecurity and the Famine of 1945

It was this agrarian setting -- combined with the reliance on the widely fluctuating rainfall for growing cereals -- that determined employment opportunities, and generated conditions of poverty and food insecurity in rural Morocco, particularly in the South. Only 10 percent of total arable land, mostly in the North and the West, was irrigated in 1950, growing not cereals, but the high-value cotton, citrus and vegetables. During the period 1945-1950, cereals occupied 60 percent of total arable land, whereby wheat and barley were cultivated in 81 percent of total cereal area. During the prolonged drought of 1945, cereal reserve was depleted, and a disastrous famine killed half the total livestock and unspecified number of poor *fellaheen* in rainfed lands. The data given in Table 5.1 indicate that yields of wheat fell to 170 kilograms per hectare on average, compared to 490 Kilograms in 1934-1938, and 587 in 1947-1950. The total harvest of cereals was only one-quarter of its average in normal times (1947-1950). As documented by Nouvele (1949) and Swearingen, (1988, p. 122), starved and distressfull small farmers sold their lands to merchants and fled to cities. This human disaster led to issuing the Law of 1945, prohibiting any sale or mortgaging of lands below 7 ha. rainfed and 1.5 ha. irrigated.

TABLE 5.1 Production of Wheat and Barley before and after the 1945 Famine in Morocco, 1934-1950

Year	area (thousand hectares)	Wheat yield (Kg./ha)	production (thousand tons)	area (thousand hectares)	Barley yield (Kg./ha)	production (thousand tons)
Average 1934-1938	1,283	490	631	1,716	670	1,148
1945	925	170	156	1,365	120	167
1947	1,150	470	542	1,431	790	1,131
1948	986	690	680	1,537	910	1,431
1949	930	590	550	1,640	610	1,021
1950	1,259	600	755	1,962	550	1,075

Source: *Production Yearbook,* Vol. I, IV and V, FAO, Rome

Partial Agrarian Reform

Like Tunisia, but unlike Algeria, the repossession of foreign-owned land was not of immediate priority in Moroccan post-independence rural development strategy. Instead, priority was given between 1956 and 1960 to the expropriation of the relatively small area owned by the Moroccans who collaborated with the former French administration, and those who opposed the activities of the Liberation Army. The area affected was nearly 12,000 hectares or 0.2 percent of total area of landholdings. Pascon's study of the Haouz Region of Marrakesh indicates that only 19 percent of the area seized was redistributed to tenants who were actually cultivating the lands, while 81 percent was retained by the State and leased-out through public auction (Pascon, 1986, Table 3.1).

It was only in September 1963 (eight years after independence), that the decolonization of foreign-owned land was initiated by *al-Dahir al-Sharif* (decree) and was included in the First Development Plan, 1960-1964. An area of 250,000 hectares of official colonial lands was gradually acquired by the State through land reform Laws of 1963, 1966 and 1973. However, after the declaration of independence, the sale of *privately* colonized land was bilaterally arranged between individual foreign-owners and Moroccans, including senior government officials and city merchants. The foreigners' unsold privately-owned land, together with the area of ex-French official colonization, amounting to 740,000 ha., was acquired by the State.

According to the 1986 report of the Ministry of Agriculture and Agrarian Reform (*al-Felaha fi tanmiyya*, p. 43), only 327,008 hectares were slowly redistributed to 23,600 families between 1966 and 1985. The rest (about 440,000 hectares) was retained by the government and managed as state farms. This means that the number of beneficiaries represented only 1.6 percent of total agricultural households. The size of distributed units varied according to the productive quality of land, ranging from 5 hectares of irrigated land to 16-23 hectares rainfed per household. Approximately, one-quarter of distributed units were in irrigated zones. As followed in Algeria and Tunisia, the criterion was a minimum annual gross income per household. On average, the value of the produce from a distributed unit was 4,000 dirhams a year (5 dirhams equalled one US dollar at the official exchange rate). This annual income level corresponds to the overall average (not minimum) expenditure per rural household, estimated by the national household expenditure survey in 1970/71. It was reported that the average level of the sale price was half the market price payable over 30 years in annual installments.

Thus by 1985, the State has become the largest single landowner in the country, owning about 440,000 ha. or 6.5 percent of total cultivable land. In addition, the State owns forest lands and most of range lands, and holds property rights *(raqaba)* of nearly 1.5 million ha. of the tribal lands. Clearly and despite the uncertain statistics, there was no shortage of productive land at the disposal of the State for redistribution to landless peasants if there were a firm government commitment to provide greater access to land, and to reduce inequality of rural income distribution as a priority in rural development strategy.

Furthermore, in spite of the post-independence hints, land policy has, since then, excluded the Moroccans' private large farms from the redistributive program.

TABLE 5.2   Distribution of Land Holdings by Size in Morocco, 1974

Size (in hectares)	Number	Percentage	Area (in hectares)	Percentage
Less than 5	1,079,090	73.6	1,771,900	24.5
5-10	219,790	14.9	1,507,200	20.8
10-20	114,050	7.8	1,525,200	21.0
20-50	43,840	3.0	1,215,300	16.7
50-100	7,720	0.5	512,300	7.0
100 and over	2,520	0.2	699,500	10.0
Total	1,467,000	100.0	7,231,400	100.0

Gini coefficient of land concentration 0.755

Source: Ministry of Agriculture and Agrarian Reform, *Recensement Agricole*, 1973/74, in Arabic. Percentage errors in the Ministry's Table are corrected. The Gini index is calculated by the author.

Available data on the distribution of *Melk* (private) lands (excluding land owned by foreigners, *habous*, and tribal lands) show that, at the time of declaring the 1963 and 1973 land reforms, 69 percent of total agricultural households were landless farmers and owners of less than 2 hectares (Griffin, 1981, Table 2.4). Most of them were living in absolute poverty, particularly in the South where dry farming and pastoralism prevail. In 1975, the World Bank estimated that 45 percent of rural population were living below the poverty line. We shall have more to say on the incidence of poverty in Chapter 7.

Who Benefits from Rural Development Strategy?

Table 5.2 shows that after instituting a series of partial but politically publicized reforms, concentration of land remained high at 0.76 index of inequality (the Gini coefficient). In 1974, there was 10 percent of land area in the top size group of 100 hectares and more that was held by only 0.2 percent of total land holders with an average 278 hectares each. We do not know how much of this category is privately owned and how much is state farms. Nor do we know how much of these holdings is rainfed and irrigated. However, Swearingen reports that in 1986, ''of the present 625,000 hectares of modern irrigated land located in the major irrigation perimeters, large landowners own or are acquiring some 500,000 hectares'' (1988, pp. 179-80). In the bottom size category of less than 5 hectares, about 74 percent of landholders held 25 percent of total area, with an average 1.6 hectares each. Moreover, these small farms are fragmented into 5-7 parcels of roughly 0.5 hectares each.

The situation has slightly improved after the completion of the redistributive program, as suggested by the results of the 1982/83 survey conducted by the Ministry of Agriculture. However, the share of holdings in the size group of 50 hectares and over had slightly increased in number and area. So did the middle size group of 20-50 hectares. Measured in terms of the Gini coefficient, inequality was reduced from 0.76 in 1974 to 0.69 in 1982/83. Thus, land distribution has remained grossly unequal despite the *fellaheen* high expectations fostered by government's promise, political unrest (1965-1971) and a persistent demand for a wider redistribution of land.[3]

Inequality of land distribution is higher in irrigated areas than at the national level, as revealed by a study conducted by Daden (1978) in the Gharb zone during the period 1972-1976. The findings show that rich farmers (50 hectares and over), representing only 0.5 percent of total number in the Gharb, captured disproportionately about 43 percent of total government subsidized inputs. Their share in total subsidized tractors and combine harvesters reached 85 percent. We should note that tractors were subsidized at one-third their market value. This substantial transfer of public funds to rich farmers occurred also in the case of subsidized water charges that were 60-80 percent below maintenance costs. The indiscriminate provision of subsidized inputs, regardless of the size of farms, favored most large farmers, who ''should not continue to benefit from these subsidies without fully paying for what they receive'' (*World Bank Country Study*, 1981, p. ii and p. 166). The same study suggests that the government's investment preference for irrigated, over rainfed, areas was contrary to the World Bank's findings that the internal rate of return on investments in rainfed areas is superior to that on irrigated lands by a high margin of 50 to 100 percent.

To understand the implications of this policy for equity, rural welfare and food security, we need to take into account the following agrarian characteristics.

1. The limited scope of redistributive agrarian reforms by which only 2 percent of total agricultural householdds were land recipients.
2. The high degree of land concentration.
3. The high percentage of pure landless households estimated at 33 percent of total rural families (Ennaji and Pascon, FAO, 1988).
4. The limited share of agriculture (10-13%) in total institutional credit supply which benefited about 30 percent of total landowners, mostly medium and large farmers *(Caisse National de Crédit Agricole* and *Banque du Maroc).*

Until 1985 when a trade liberalization policy was introduced, those who benefited most from government subsidies and pricing policy were the farmers in irrigated areas who have already obtained numerous advantages noted earlier. Apart from their land-based power, they also have secure access to perennial water supply. This does not mean that farmers in rainfed agriculture did not receive benefits from public actions. In fact, a number of integrated rural development schemes were implemented, especially in the Haouz region and the middle planes of the Atlas agricultural regions. Although they were not specifically targeted to benefit the poor, these schemes provided farmers with small-scale irrigation equipments, and they improved health and education services. Integrated rural development efforts have also improved pasture lands and animal husbandry in rainfed areas *(al Felaha fi Tanmiyya Mustamera*, Ministry of Agriculture, 1986, pp 34-40).

Policy Impact on Food Production

Considering that approximately 60 percent of total cultivable land grow wheat which constitutes on average two-thirds of the consumer's calorie-intake, wheat price policy has influenced both land allocation among crops and food security. During the period 1970-1984, prices of cereals were controlled by the government, thereby the producer's price was kept 40-50 percent below its effective market price. Apart from high instability of rainfall, a combination of land tenure arrangements, crop pricing and subsidization policy in the 1970s had their effects on farmers' incentives and food production between 1970 and 1980. The average annual rate of food production growth was very low at only one percent, while population grew at the higher rate of 2.4 percent, the inevitable result was a fall in food production per head to a negative 1.4 percent and a sharp rise in total food imports by 59 percent. Wheat production, in which Morocco has a comparative advantage stagnated, and its import doubled during the same period (FAO *Country Tables*, 1989). Morocco also received a substantial amount of wheat aid with an annual average of 190,000 tons during the 1970s, reaching 0.45 million tons between 1983 and 1984/85 (*Food Aid in Figures*, 1985 and 1987).

TABLE 5.3 Changes in Wheat Production and Prices in Morocco, 1965-1989

	Average 1965-1969	Average 1979-1983	1985	1987	1989
Area of wheat (thousand ha.)	1,895	1,712	1,895	2,288	2,630
Yield (ton/ha.)	0.87	1.01	1.25	1.06	1.49
Producer price (Dh per ton)	n.a.	125	180	200	220
Wheat production index	100	105	144	147	238

Notes: Dh is Moroccan dirham. n.a. stands for not available.
Sources: Area, production and yield from *Production Yearbook*, several issues, FAO, Rome. Producer price is from the Ministry of Agriculture, Rabat.

Starting 1985, the government implemented the IMF and World Bank-induced program for trade liberalization and structural adjustment, including a 40 percent rise in producers' prices and a gradual elimination of input subsidies (Ministry of Agriculture, 1990). Table 5.3 suggests that price policy has provided producers with incentives to increase the 1989 wheat production to more than double its volume in 1979-1983. We should not overlook the production effect of the expansion in irrigated land by one-third between 1970 and 1990.

This substantial rise in harvested area by 54 percent and producer's price by 76 percent must have also benefited small farmers in the size category of less than 5 hectares who represented 68 percent of total landholders. They grow wheat in almost half of their holdings' area. Furthermore, as most of rural migrants to urban areas were from this category of small farmers and the landless wage workers, their food security and average annual income must have been improved. We shall have the occasion to say more on equity, standard of life and poverty changes in Chapter 7.

Libya

Unlike the rest of the Maghreb, Libya (al-Jamahiriyya al-Libiyya), has a small population (4.5 million in 1990), growing at the very high annual rate of 4.3 percent. It has also the lowest density of agricultural labor force on arable land in the Maghreb, only 0.07 person per one hectare. But like the other Maghreb countries, Libya was ruled by the Italians (1911-1942) and jointly by the British and the French between 1942 and its independence in December 1951. Important among the agro-ecological common features is the reliance of cereal producers and pastoral nomads on the unstable rainfall.

The discovery of rich oil deposits in 1959 was a significant inducement to far-reaching changes in the Libyan economy. A single but internationally comparable indicator of these changes is the *per capita* income. In his study of the Libyan economy in 1952, Higgins ranked Libya among the poor underdeveloped countries, and estimated its annual *per capita* income at US $50. In 1985, Libya was classified, by the World Bank Development Indicators, among the high-income countries, having $7,170 GNP *per capita*.[4] Using a broader measurement of development, the 1991 United Nations' Human Development Index for Libya was the highest among the North African countries. The implications of this dramatic change for agriculture and the rural people constitute the subject of this section, after understanding the situation in the 1950s.

The Tribal Economy before the Oil Boom

The Tribal System in Agriculture

The tribal system and its communal rights in land was predominant in rural Libya. It had combined deeply rooted customs and agro-climatic requirements, and it was able to meet the country' food needs, as well as to employ a major section of rural population. For nearly four centuries, it had resisted agents of change during colonial rule. Throughout their long rule, the Ottoman Turks did not succeed in the imposition of the Ottoman Land Code of 1858, requiring the registration of *individual* land use rights for taxation purposes. Thanks to the persistent opposition of the tribal leaders, the customary-tribal system survived also the Italian rule which had the clear aim of aggressively settling the immigrant poor Italian peasants in the fertile land of the Mediterranean coastal zone, after dispossessing its native holders. During the short period of the joint British and French administration immediately before independence, the customary agrarian institutions remained unchanged. It was only in the 1960s and 1970s that this rural institution started to give way to the impulse of technical and economic changes brought about by the oil boom and the abolition of tribal communal rights in land by state intervention.

According to the results of 1959/60 census of agriculture, the area of land held communally by the influential tribes amounted to 40 percent of total cultivable land plus pasture land. The rest was held by the former Italian settlers and business men, both Libyans and foreigners. Only 1.2 percent of total area of Libya was arable and woodland, out of which nearly 5 percent was irrigated, and 85 percent rainfed. The rest, or approximately one-tenth, was shrubs and forest areas. Out of cropland, cereals and peanuts were predominant and only 3-4 percent grew, in 1959/60, cash crops (fruit trees, vegetables, tobacco, vineyard and sugar beets). During the Italian rule, about 12 percent of land growing cereals was converted to tobacco and vineyard cultivation; the latter was for making wine, which was against Islamic principles prohibiting drinking alcohol.

Dual Agrarian Structure

The above paragraph suggests a dual production structure consisting of two sectors having distinct features. The modern and commercialized sector comprized large farms, mostly mechanized and owned by foreigners, including ex-Italian settlers, the average size of which was 40 ha.. These farms were situated in the high rainfall zone (400-600 mm *per annum*), irrigated in the dry season, and produced olive cash crops and cereals. On the other hand, the traditional sector embraced the bulk of the agricultural population. Given the dependence of two-thirds of total population on agriculture, pastoral nomads represented 57 percent of agricultural population in 1960 and the rest were wage-dependent workers in the modern sector, and holders of usufruct rights in land belonging to several tribal organizations. Landholdings were fragmented through inheritance arrangements into 7 parcels on average of 2 ha each. Small farmers grew cereals and raised sheep, and they were vulnerable to sandstorms and rainfall hazards. Those who irrigated small parcels for growing vegetables manually raised underground water from shallow wells in buckets *(dalw)* in contrast to sprinkling irrigation and large size deisel and electric pumps in the modern sector. We should note that Libya has no natural rivers but has several wadis and the man-made river which is still under construction.

Thus a wide disparity prevailed in the distribution of wealth, productivity gains and income; so did the quality of life in terms of education, health and housing.[5] This gross inequality in land, opportunities and rewards was rooted in the land tenure system, cropping patterns and the neglect of the vast traditional sector during the long colonial rule.

Rural Development in an Oil-based Economy

Ideological Shift

The sequential events in the 1960s and 1970s suggest two principal factors which have significantly influenced the content and pace of policy change. The first is the rapid flow of oil export revenue from US$60 thousand in 1960 to $30 million in 1963, reaching $23 billion in 1980. This rapid flow has been combined with the steadily increasing state monopoly power in the oil industry after its nationalization in the 1970s. Oil contributed about 68 percent of GNP and 99 percent of exports in the 1970s. The second principal factor is the sudden shift in the political power structure away from a coalition of tribal chiefs, rich urban traders and foreign interests, towards an equity-orientated development strategy, following the 1969 *coup d'état* led by al-Quddafi. The style and impulse of policy change seem to reflect a combination of rising nationalism, a form of socialism and reformation. Socialism meant common ownership, abolition of exploitation, according to Islamic principles, and popular participation in policy formulation. The author witnessed the shift in rural development policy and participated in its formulation in an advisory capacity.[6]

Redistribution of Landed Property

State intervention to redistribute land from certain categories of owners to the intended beneficiaries among rural people has taken place in varied forms and stages. The first and most publicized is the repossession of the ex-Italian farms. They were part of an ambitious colonization program started in the 1920s by the Italian government's *"Ente per la colonizzazione della Libia"*. However, its completion was disrupted by the outbreak of the second World War in 1939 which resulted in the defeat of Italy and the departure of the settlers. Yet, about 210 thousand ha of fertile land along the Mediterranean coast or 11 percent of cultivable land was already settled by Italian farmers. The dispossessed tribes were not compensated, and were expelled south.

Two paths were followed for the acquisition of these farms. The first was through the market, whereby some private Italians sold out their farms in the early 1960s. The exact area of this private transaction is not known. The second form of recovering the Italian landed property was through a negotiated agreement in 1956 between the Libyan and Italian governments amounting to 25,000 hectares. By force of a series of laws issued in 1970, the remaining land privately owned by Italians, together with other foreign-owned farms, were repossessed by the State.

Likewise, all properties of the deposed monarch (King Mohammad Idris al-Senousi) and those belonging to Libyan families identified as anti-revolution were confiscated. The 1970 laws also prohibited the ownership of land by foreigners. All requisitioned lands (about 115 thousand hectares) had to be developed for redistribution and for retention as state farms, their irrigation during the dry season ensured, and their housing facilities either rehabilitated or newly constructed by public funds. Once these works were completed, land, houses, stores and equipment were distributed among tenants, wage workers and the descendants of the dispossessed tribes, if they were still farmers. Land and capital equipment were sold to the beneficiaries at half the estimated value, while houses and stores were granted free of charge. Depending upon soil fertility, the size of distributed farms ranged from 2-8 ha. of fully or partially irrigated land to 40-60 ha. rainfed, or 20 ha. of mixed.

Directing Oil Revenue for Rural Development

Post-1963 development plans had aimed to reduce the dependence of the national economy on a single commodity, oil, to attain food self-sufficiency and to check rural migration to urban centers. Since the pursuit of the land reform policy outlined above, varying shares of oil export revenue have been allocated to improve the appropriated land, subsidize agricultural production change, expand the area of cultivable land, and to improve the quality of life in rural areas. Following the recovery of most of the ex-Italian farms and the expropriation of affected land, an amount equivalent to US$34 million was allocated from 1964 to 1969 for their comprehensive development.[7] This amount was for land reclamation, irrigation, rural roads, housing and administrative costs. The share of agriculture in development expenditure was 10 percent during the same period.

To develop rural Libya, heavy public investment was required to bring ground water to the surface, control the seasonal floods of the *Wadis* and to prevent the waste of rain water flowing into the Mediterranean. Complementary works included land levelling, and the construction of irrigation canals and drainage facilities. In addition, rural infrastructure (roads, houses, stores, safe drinking water, schools, clinics and electricity) were provided free of charge to serve the new land recipients and the local staff.

Expanding Cultivable Land. Between 1970 and 1980, it was planned to reclaim about 1.2 million acres for distribution among eligible farmers, college graduates and public corporations responsible for food supply and livestock production. Out of the total area planned, 300 thousand acres were to be irrigated and the rest rainfed.[8]

This ambitious program is in addition to the project known in Libya as "the great man-made river" which has, since August 1984, been under construction by a South Korean firm for irrigating, by the year 2000, nearly half a million hectares in the Eastern region plains between Benghazi and Sert. The first phase of the scheme - costing nearly ten billion US dollars - provides for drilling water (from deep layers of the aquifer in the coastal areas and wells in Kufra oasis in the South) and distributing it in huge pipelines. Given the current weakness in enterpreneurship and organization, it remains to be seen in the year 2000, whether this grandiose scheme would be a miraculous achievement or a mirage in the Libyan desert.

Subsidization Benefits to Farmers. Another form of directing oil revenues to benefit farmers is subsidization. Prices of agricultural production inputs needed by all farmers are subsidized at rates, ranging from 50 to 80 percent. They are provided by the State-owned Agricultural Credit Bank through a network of state-patronized co-operatives, membership to which is obligatory. Farmers and drillers of ground-water wells receive 50 percent subsidy and interest-free loans for the purchase of pumps, windmills, and related equipment, payable in annual instalments over 15 years. The average annual total subsidies enjoyed by farmers in the 1970s was half a million Libyan dinars or US$ 1.65 million at the official rate of exchange (one dinar equalled 3.3 US dollars).[9]

The source of funding subsidies and land reclamation was the development budget, which came from oil and natural gas export revenues. This budget is distinct from the ordinary or administrative budget which is predominantly funded from non-oil sources. The share of agriculture in the former was raised from 10 percent in the 1963-69 Plan to 13 percent in the 1969-1974 Development Plan, and it was raised further to 20.4 percent in the 1976-1980 Plan. However, the share fell to 12 percent in 1986-1988 according to the Libyan Central Bank. This fall was mostly due to external forces, including the collapse of world oil prices, the world oil glut, the imposition of trade sanctions by the U.S. Government, and the military expenditure required during the period 1986-1988 for the Libyan War against Chad. Between 1980 and 1989, oil export revenue fell sharply from the equivalent of US$ 23 billion in 1980 to US$ 5 billion in 1986, and it recovered slightly to $ 7.5 billion in 1989. Despite this plunge in oil revenue, the shares of both health and education services have not been reduced. Rather, they have been sustained throughout the 1980s at 10 and 14 percent, respectively.

The Impact on Rural Development: An Assessment

One would expect that all public actions described earlier should have, in varying degrees, generated agricultural growth and improved income distribution. It is also likely that the improvement in the distribution of income within the rural sector has induced private investment to raise output, through the purchase of chemical fertilizers, tractors, electric pumps for irrigation, and improved seeds and livestock breeds. The number of tractors in use doubled between 1975 and 1985, and the consumption of fertilizers increased during the same period by 150 percent in terms of kilograms per hectare of cultivable land (*Country Tables* and *Production Yearbook*, FAO 1986).

High Social Cost

It does seem that economic considerations of resource-use efficiency, cost-effectiveness and economic return on invested oil revenue in agriculture and, especially in land settlement schemes were of less concern to policy makers than the realization of quick political gain. One explanation of this order of preference is the availability of plentiful, sudden, and rapid flowing oil revenue, while the availability of development-adapted leadership was scarce in the agricultural sector. Another possible explanation was the high cost of national unity required by the recent emergence of independent Libya out of long colonial rule and being formed of three large regions, previously administered separately and suffered from tribal rivalry (former Fezzan in the South, Tripolitania in the West and Cyrenaica in the East). They were also separated by hundreds of kilometers of desert.

In the process of redistributing wealth for rural development, the policy makers have tended to overlook the institutional constraints: complexity of social change, the weak administrative capacity of state institutions, and the scarcity of farming experience among the intended beneficiaries. It was learnt from bitter experience that the lavish public spending to modernize agriculture could not quickly transform a tribal-based community and its subsistence farmers, having recent nomadic background, into capital intensive and technology-based entrepreneurs. Likewise, the mere issue of a complex body of legislation could not speedily convert most of the rural population, having traditional allegiance to the tribal and sub-tribal chiefs, into farmers obeying directives from local officials and managers of the imposed co-operatives. Hence the inhibited producers' motivation.

However, a slow progress was made. Tangible technical change in the old farming areas was realized after one decade of persistent government efforts, continuing flow of oil money, and hiring a large number of highly paid foreign experts specialized in all aspects of agriculture. The technological change was manifested in the adoption of improved seeds, the accelerated consumption of chemical fertilizer, the expansion in irrigated area, and the production of high value crops and vegetables (see Table 7.1). It also revealed itself in the notable increase in animal production together with improved veterinary services on farms, and in fixing wire fences around vegetable and alfalfa fields.

Considering the eventual parcelation of land from one generation to the next, and given the abundance of capital, expediency determined the smallness of the size of distributed farms in order to meet immediate demands and not a perspective long-term development. According to the findings of a sample survey of 371 farms conducted in the Western region in 1967/68, the majority of farms with irrigation facilities (84%) were less than 5 hectares (12 acres). Only 6 percent of old farms and 35 percent of the newly developed were in the size category of 5-25 ha.[10] These sizes were almost half those recommended for capital intensive farming in a land and capital-plentiful Libya. The smallness of the units are likely to lead to diseconomies of scale.

Moreover, the social cost of land settlement schemes was high. Based on data calculated by the Ministry of Planning in 1975, the development expenditure (from oil export revenue) per household and per hectare in a sample of five resettlement schemes is very high and varies widely from US$ 81,000 to US$ 210,000 per household settler. An explanation provided by El-Wifaty (1978) for the high cost is summarized as follows:

1. Political pressure applied on the implementing departments for rapid completion of the schemes.
2. Diffused responsibilities and accountability among several government departments in charge of land reclamation, irrigation and drainage works, rural housing and electricity supply.
3. Speedy importation from Europe of expensive building material and irrigation equipment to meet the deadline set by policy-makers.

4. Contracting out work to internationally reputable firms, charging high fees and whose work in the implementation of schemes was not adequately monitored by government departments, due to their lack of qualified staff.

Effects on Agricultural Production and Food Security

The production performance of agriculture indicates a steady growth of agricultural GDP between 1960 and 1988. Growth was slow at an average annual rate of 3.5 percent in the 1960s, then accelerated at 11 percent in the 1970s, but fell slightly to 10.7 percent during the 1980-1988 period. These high rates of growth of product are valued at fixed prices by the World Bank. The data on the growth of agriculture products calculated by FAO also show an upward trend but at lower annual rates: 7 percent in 1971-1980 and 3 percent in 1981-1988.[11] The rise in production was faster in the livestock sub-sector than in cereals. Consequently, cereal imports rose by 58 percent between 1980 and 1988 and by 134 percent between average 1974-1975 and 1988. Moreover, total food imports as a percentage of total domestic food consumption increased from 69 in 1981 to 78 in 1988.

Recently, CIMMYT estimated that, in 1987-1990 *per capita* annual cereal (wheat and barley) production was 70 Kilograms and consumption 340 Kilograms. These figures suggest that production was insufficient by 79 percent of consumption requirements (1990-1991 *World Wheat Facts and Trends.* p.35). This approximate magnitude of insufficiency is in sharp contrast to the Ministry of Planning's target of 91 percent self-sufficiency in cereal production in 1985 (*Summary of the* 1981-1985 *Development Plan,* Table 5).

Thus, Libya failed to realize its development objective of self-sufficiency in food from domestic production. This increasing reliance on food imports is partly due to the rise in food demand by the fast growing population, and partly because of the fact that food (particularly wheat) production has not increased fast enough to outpace population growth, resulting in falling productivity per person. Average annual rate of growth of food productivity fell from 1.7 percent in the 1970s to a negative 1.7 in 1980-1988, leading to a sharp rise in food imports, especially wheat, both grain and flour. One is inclined to doubt the accuracy of official statistics on agricultural growth which should be treated with caution.

Undoubtedly, directing oil revenue to expand cultivable land, and to intensify its use through irrigation investment has substantially contributed to agricultural growth.[12] Although the actual expansion in irrigated area has been far below the planned targets, it increased by 93 percent from 121,000 ha. in 1960 to 234,000 ha. in 1985. It appears that the expansion is positively associated with the changes in the share of agriculture in development expenditure as suggested by the data given in Table 5.4.

TABLE 5.4 Investment in Expanding Irrigated Land in Libya, 1964-1989

	1964-1970	*1970-1980*	*1980-1985*
Average share of agriculture in total development expenditure[a]	10%	17%	13%
Planned increase in irrigated land in thousand hectares[a]	n.a.	100	33
Actual increase in irrigated land[a]	40	55	14
Gross fixed capital investment in US $[b]		*1978*	*1985*
per hectare arable land		243.7	400.0
per working person in agriculture		4,463.6	6,197.8

Notes: Gross fixed capital formation (or investment) is a valuation of investment in land improvement and reclamation, irrigation, drainage, farm buildings and equipment.
Sources: a. Libya Development Plans and the *Central Bank Bulletin,* several issues. b. United Nations' *National Acount Statistics,* calculated per hectare and per working person by FAO and published in *The State of Food and Agriculture,* (SOFA), 1981 and 1989, (Annex Tables), FAO, Rome.

We should note that, in practice, there is a time lag of 5-8 years between actual capital expenditure for land development and the realized output growth. Apart from the delay caused by weakness in coordination among the several government departments (responsible for land reclamation, irrigation, cultivation and road construction), different physical characteristics of soil require different periods for bringing land productivity to the desired level. Moreover, the multiple effects of capital expenditure on production growth do not occur once at a point in time. Rather, they are brought about during the lifetime of agricultural development activity, particularly with continuing private and public investments.

The Impact on Employment and Farmers' Incomes

In the 1960s, the low productivity and earnings of agricultural population combined with a slow growth of agricultural output induced rural-urban migration. At that time, the impact of oil was widespread in urban areas where rural migrants could find easy and lucrative earnings, particularly in the construction boom, government services and trade. Consequently, urban population grew swiftly at the average annual rate of 8 percent between 1960 and 1970, peaking in the 1970s at the rate of 9.8 percent, then slowing down at 6.3 percent in the 1980s. At least half of this growth can be attributed to rural migration. Concurrently, the size of agricultural population fell by 22 percent and agricultural labor force by 29 percent (1960-1980).[13] Thus, the aim to check out-migration by heavy public investment in agriculture has not been realized. In the already sparsely populated rural Libya, the agricultural sector had lost nearly 58,000 working persons, and rural population had diminished by 0.4 million.

At the same time, the rapid expansion of compulsory free schooling in rural areas has contributed to the decline in native labor supply in agriculture. Induced by attractive wages and working conditions in urban activities and government jobs, the young and educated rural people were the first to abandon agriculture, leaving behind the elder and female population. The influx of cheap migrant workers from Egypt and Tunisia filled partially the created gap in agricultural labor requirements. We say partial, because 9 percent of rural population in 1973 were foreign workers according to the 1973 population census, while the Libyan people working in agriculture had diminished by 29 percent between 1960 and 1975. Hence the reduction in total supply of labor and the rise in wages.

TABLE 5.5 Rising Income of Working People in Libyan Agriculture, 1972-1980

	1972	1975	1980
1. Average income (GDP) per person of total population, L.D.	620	1412	1940
2. Average agricultural GDP per working person in agric. L.D.	406	930	2006
2 as percentage of 1	65	66	103

Notes: L.D. = Libyan dinar at 1974 fixed prices. It equalled US $ 3.3.
Source: Calculated from the Ministry of Planning data on gross domestic product and the shares of all sectors which are outlined in the 1973-1975 Development Plan and 1976-1980 Socio-economic Transformation Plan. Data on labor force in agriculture are from *Country Tables* 1990, FAO, Rome.

In the face of these changes in the labor market and the rapid expansion in government jobs, small landowners responded to these new economic opportunities in different ways. Some became absentee, sub-contracting their farms to expatriate workers and took up government jobs. The State has virtually replaced the market as the determinant of wage rates and employment opportunities. By 1975, only 55 percent of total landholders were full-time farmers (owners-operators), while the rest (45%) were part-time farmers and absentee landowners, earning most of their income from the less productive government jobs and other non-farm activities.[14]

Undoubtedly, those who remained on the land gained from the availability of cheap foreign labor, government price support of crops and the generous subsidization of the means of production. They greatly benefited from the fast-rising demand for food consumption in urban areas. Table 5.5 shows the sharp rise in average real income of the working people in agriculture from 1972 to 1980.

Based on the estimates given in Table 5.5, the income gap, between that of a person employed in agriculture and the national average, was narrowed during the 1972-1980 period. The rise of the former was much faster than the latter in 1980, suggesting a substantial rise in *per capita* real income in agriculture, compared to the national average. However, with the fall in growth rates of agriculture during the period 1980-1985, noted earlier, the gap widened. Using the World Bank's internationally comparable data on *per capita* income in US dollars, the average income per working person in agriculture as a percentage of that *per capita* total population, fell in 1985 to 68.

Nevertheless, the average income level and the annual growth rate of agriculture in Libya remain quite high by developing country standards. Moreover, the rise in real income in agriculture, combined with sustained and adequate allocation of oil revenues to health, education, safe drinking water and housing in the 1970s and the 1980s, must have considerably raised the level of living in rural Libya from its appallingly backward conditions in the 1950s. Yet the planners' obsession with food self-sufficiency from domestic production, and checking rural out-migration was unrealistic, and did not work.

Notes

1. See Lazarev in Etienne, ed., (1977).

2. This statement and the discussion that follows are based on an unpublished study which I prepared in 1989 for the United Nations ECA. The material is used here by permission of the Joint FAO/ECA Agriculture Division, Addis Ababa.

3. Using the official data on the distribution of land and those of the United States' Agency for International Development (USAID) unpublished report of 1986, "Morocco, Country Development Strategy", Swearingen shows how the pace of agrarian reform was associated with the Casablanca riots in March 1965, and the attempted *coup* by rebel military officers in July 1971 and August 1972 (1988: Table 13 and p. 177).

4. Income in 1952 was estimated by Benjamin Higgins, *'Economic Development: Problems, Principles and Policies'* (New York: 1959, p. 26). Income in 1985 is from *World Development Report, 1987,* Development Indicators, Table 1. For the Human Development Index, see note 16 and Table 7.5 in Chapter 7.

5. For a detailed description of the rural setting in the 1950's, see El-Ghonemy, "The development of Tribal Lands and Settlements in Libya" *Land Reform,* FAO Journal, Rome, 1965. See also *Land Policy in the Near East,* El-Ghonemy, ed., published by FAO and the Government of Libya (Rome: 1967), and the World Bank, *The Economic Development of Libya,* 1960.

6. Following my study in 1961, the Government of Libya asked me to prepare a project "The Development of Tribal Lands and Settlement of ex-Italian farms" which was implemented by FAO and financed by the Libyan Government. This project ended in 1970, of which the writer was advisor. Between 1970 and 1977, the writer was frequently advizing the Government on land redistribution program and rural development policy. On the new ideology, see Muammar al-Qaddafi, *The Green Book*, 1976, in Arabic and English.

7. For a detailed account of the program implemented in the 1960s and public funds allocated, see El-Jawhary in El-Ghonemy (editor), 1967 and the report of FAO on its Fund-in-Trust project in Libya "*The Development of Tribal Lands and Settlement - Summary and Recommendations*", Rome, 1970. See also El-Ghonemy (1965). The Libyan pound renamed dinar equalled US$ 3.3

8. The data on the planned program are taken from the Ministry of Planning records for the agricultural sector. See note 9 below.

9. See the study prepared by Secretariat of Agriculture, 1978, in Arabic.

10. *Libya: Agriculture and Economic Development*, edited by J.A. Allen, K.S. McLachlan and Edith T. Penrose, Table 4.10. This study was jointly carried out by the University of London and the University of Tripoli, Libya.

11. GDP growth rates are from *World Development Report*, several issues, and agricultural production growth rates are from FAO *Country Tables*.

12. According to the Ministry of Planning and the Central Bank annual reports, agriculture contributed to gross domestic product, at current factor cost, 315 million Libyan pounds (dinar) on average between 1985 and 1987, while food imports absorbed, during the same period, an average amount of 256 million dinars.

13. Urban population expansion is from *World Development Report*, several issues and the results of the Population Census of Libya, 1973. Changes in growth rates of agricultural population and labor force are calculated from FAO *Production Yearbook*, several volumes.

14. See source in note 9. These figures are cited in p. 68 of the study.

6

Food Insecurity in the Nile Valley: Egypt and Sudan

The grouping of Egypt and Sudan should not be interpreted as having a political meaning. Apart from their historical-political linkage, up to the independence of Sudan in 1956, their economy and the well-being of a large section of their population depend on the Nile water, the quota of which was allocated by mutual agreement in 1959. Likewise, as we shall soon find out, the technical change through irrigation has fundamentally changed their agrarian institutions, land profitability, foreign trade and resource use, particularly after the expansion of cotton production.

Nevertheless, significant variations do exist in agrarian structure. They include the system of land tenure, the density of labor on cultivable land, the intensity of land use, the degree of cereal production instability and the extent of potential cultivable land. There is also an important variation in the proportion of settled and nomadic population. The density varies widely; in Egypt it is seven times that of Sudan. Whereas Sudan possesses the largest potentially cultivable area in Africa (after Zaire), Egypt's potential is very limited. Available estimates suggest that Sudan's potential is more than three times the present cultivated land, while it is only less than one-fifth in Egypt; the constraint being water and not land. Seemingly, the density factor and the historical evolution of social structure have influenced each country's rural development policy. Their policies are distinctly opposite in content.

Despite the flow of water secured by the Nile endowment and its valley's fertile soil, both countries have been increasingly unable to feed their people from domestic food production. This is due partly to the fast rising aggregate food demand, resulting from high rates of population growth and rapid urbanization, and partly to mismanagement of land and water resources. In Sudan, the food problem is compounded by its dependence for cereal production upon the widely fluctuating rainfall and by prolonged droughts. One result of these factors has been a declining average annual growth rate *per capita* food production. The other result has been the persistent rise in food imports (including aid), taking up an increasing proportion of agricultural export earnings.

Having introduced the primary features of the rural economy in land-scarce Egypt and land-abundant Sudan, the discussion turns to explore, in an historical context, the past factors that have shaped their current agrarian systems, and recent policies influencing food insecurity and levels of living in rural areas.

Egypt

The evolution of land policy which has, since 1805, cumulatively shaped the present rural economy is reviewed before analyzing its effects on food production and income distribution. While the discussion covers a long span of time, it places special emphasis on the post-1952 period when government intervention has been extensive.

The State and Agrarian Institutions in an Historical Perspective

Pre-1952 Land Concentration and Polarization

In 1805, the ruler, Mohamed Ali, declared Egypt's break from the Ottoman Empire and nationalized land property during the period 1810-1820, deciding how much, by whom, and under what terms land should be held. Also, he controlled directly *waqf* lands amounting to nearly 600 thousand feddans (one feddan equals 1.04 acres or 0.42 hectares). His administration monopolized the marketing of wheat and cotton, and forced, for security purposes, most nomadic people (bedouins) to settle in the Nile Valley, where they reluctantly practised farming. He then granted land use rights to those influential groups on whom his power depended. Under this state monopoly, different forms of agrarian institutions (*Uhda, Shiflic, Ibadeya*) were created against payment of differential rates of land taxes.[1] Areas ranging from 500 to 8,000 feddans were granted to holders of high office in government and the army, Moslem leaders and favored Egyptian families in the countryside. According to the account given by the Egyptian historian Ali Pasha Mubarak (1887) and Baer (1962), most of the large landownerships which existed up to Gamal Abdul-Nasser's revolution in 1952 originated during that period.

Land use rights in these grants were eventually converted by the Sovereign between 1860 and 1880 into private ownership. In 1878, the ruler (Khedive

Ismail) and members of his royal family alone held among themselves nearly one-fifth of the total agricultural land, including most of the best land in the country (Baer, 1962, p. 41). In the meantime, the *fellaheen* cultivated small units of land (2-5 feddans each) against payment of land tax.

The motive of this policy was chiefly fiscal: the collection of land tax (*Ushr* and *Kharaj*) as the main source of total public revenue, representing on average 93 percent of total direct taxes between 1835 and 1881 (Owen, 1969, Table 60). Likewise, financial aims were most likely behind the conversion of the usufruct rights in land to private landownership against the advance payment of six times land tax during the country's financial crisis in the 1870s. As occurred in Tunisia, the crisis resulted from government policy to accommodate heavy public debts owed to British and French financiers that led to the British occupation of Egypt in 1882. Nearly one-tenth of total cultivable land, or 188,000 hectares, were sold for this purpose, a large part of the government-owned agricultural land. Much of this land was purchased at low prices by European land corporations which, by 1890, held 11.5 percent of total privately owned land (El-Ghonemy, 1953). Large Egypto-Turkish landlords also took advantage of the sale, purchasing nearly 300,000 feddans, thus contributing further to a rising concentration of wealth and power in rural Egypt.

To grapple with heavy public foreign debt, agricultural land taxes were sharply raised from one-fourth of the harvest to one-third, and then to one-half in 1870. The consequences were socially harmful. Small owners who were unable to pay the taxes were forced to forfeit their properties, the total area of which was estimated by different authors at 200 and 300 thousand feddans which ended up in the hands of government officials, large landowners, moneylenders and village sheikhs (Artin, 1883, and Owen, 1969, p. 148). The *fellaheen* who lost their lands migrated to towns or became landless workers, and large landowners accumulated greater wealth.

Irrigation Expansion and Rising Land Profitability. These manifestations of increasing demand for land have to be understood alongside the radical technological change in irrigation between 1840 and 1870. The Nile flood had been brought under control by irrigation engineering in order to intensify cropping and to reclaim new lands. Consequently, the ubiquitous basin irrigation (*heyad*) was virtually replaced by the tightly state-managed perennial irrigation system. Whereas the former permitted the cultivation of one crop immediately following the flood season, the new system supported two or even three crops grown in the same area in a single agricultural year, including the high value and exportable long-staple cotton. By the end of the last century, the cultivable area increased by 60 percent, total cropped area doubled and the intensity of land rose by 40 percent (see Table 6.1). Within a short space of time, there was a dramatic rise in areas of food crops and cotton (in thousand feddans based on O'Brien, 1968, Table 1) as follows.

	Wheat	Millett & Maize	Beans	Rice	Cotton
1844	914	799	839	98	699 (1871)
1878	1,150	1,900 (1879)	1,220	240 (1879)	950 (1879)
rise %	26	138	45	145	36

Accordingly, the marginal product (and value) of land rose sharply and profits from cotton determined income and wealth. Henceforth, tenancy arrangements responded to this technological change. Rental values were changed from payment in kind to cash with less sharecropping, and they have followed the movement in cotton prices (see Figure 6.1). Moreover, starting in 1899, the assessment of land tax began to be based on land rent. But despite the rise in land profitability from growing cotton, the powerful landlords in coalition with the government managed to reduce land taxes from 28.8 percent of assessed rent in 1899 to 16 percent in 1937 and reduced further to 14 percent in 1947.[2] This occurred while the average annual yield of cotton rose sharply from 3.4 kentares in 1890 to 5.2 Kentars in 1937-1939 and further to 6.1 kentars in 1948.[3]

The rising profitability of irrigated land induced private foreign capital (British and French) to invest in land ownership, reclamation and cotton trade. By 1945, foreign landowners numbered 4,570 (0.16 percent of total owners), possessed 347,000 feddans or 6 percent of total area and the average size of their farms was 2,000 feddans compared to 2.1 feddans of most Egyptian landowners (*Annuaire Statistique*, 1946-1947, p. 306). Institutional financing of agriculture, in general and land development, in particular came from three land-mortgage banks owned by British, French and Belgian investors. They monopolized the supply of agricultural credit and directed it largely to foreign land corporations and Egyptian landlords. Both foreign landowners and banks also purchased most of the reclaimed state-owned lands to the disadvantage of the *fellaheen* who were landless wage workers, sharecroppers, tenants and small owners in the size category of one hectare, growing chiefly wheat, maize, and beans for subsistence.[4] Hence, the duality of the agrarian structure took root.

The Extent of Polarization. Concentration of land ownership is manifested in the results of the 1950 census, which was the last before introducing land reform in 1952. It shows that 94 percent of total landowners were in the category of 5 feddans and less, including 71.8 percent owning less than one feddan (0.42 hectare) each. Landless wage-workers constituted between 40 and 50 percent of total rural households, depending on the definition used. At the other end of the scale, owners of 100 feddans and more, representing only 0.2 percent of total number, possessed 27 percent of total privately owned area. In the meantime, the pressure on cultivable land was mounting. Between 1900 and 1950, the agricultural population increased 110 percent, while total cultivable land increased 25 percent, resulting in a sharp fall in land per person ratio (see Table 6.1).

TABLE 6.1   Changes in the Aggregate Supply of Land, Agricultural
Population, Land Use Intensity and in Land per Capita Ratio
in Egypt, 1813-1984

Year	Cultivable Area Mill.fedd. Index (1)		Cropped Area Mill.fedd.Index (2)		Intensity Ratio (2÷1) (3)	Agric. population Million Index (4)		Cultivable land per capita (1-4) Ratio Index (5)	
1813	3.05	100	3.05	100	1.00	2.2	100	1.38	100
1835	3.50	115	-			2.6	118	1.35	98
1877	4.70	154	4.76	156	1.01	4.4	200	1.07	77
1897	4.87	160	6.70	220	1.37	6.2	282	0.78	56
1927	5.54	181	8.81	288	1.59	8.8	400	0.63	46
1937	5.28	177	8.36	275	1.58	9.3	423	0.58	42
1947	5.76	189	9.16	300	1.59	10.6	482	0.54	39
1952	5.80	190	9.30	305	1.61	13.2	600	0.44	32
1960	5.90	191	10.39	340	1.76	15.0	682	0.39	28
1965	6.04	198	10.30	340	1.71	16.1	732	0.37	27
1970	6.39	209	11.46	376	1.79	17.1	777	0.37	27
1975	6.19	203	11.16	366	1.80	17.7	804	0.35	25
1980	5.87	192	11.13	365	1.89	18.9	859	0.31	22
1982	5.83	191	10.96	360	1.88	19.3	877	0.30	22
1984	5.83	191	11.04	361	1.89	19.9	904	0.29	21

Notes: One feddan = 1.038 acres = 0.42 hectare. Cultivable area is the amount of land available for farming irrespective of its intensive use. Cropped area is the area of different crops cultivated in one feddan during the same year. Agricultural population are those adults engaged in farming as their main occupation and their dependents.

Sources: Columns 1 and 2: 1813-1897 based on data given in O'Brien 1968, Tables 3 and 10. 1927-84: Annual Statustics by Egyptian Central Statistics Office (CAPMAS) and the Ministry of Agriculture Annual Bulletin of Agricultural Economics *Al-Iqtisad al-Zira'i*, several issues in Arabic. No reliable data exists on cropped areas in 1835. Column 4: 1813, 1835 and 1877 are based on O'Brien's carefully compiled estimates on total and urban population. For 1813 and 1835 we assumed that all rural population were agricultural and that 95 percent were agricultural in 1877. From population censuses 1897 and 1927, we assumed the percentage to be 90 and 85 percent respectively. Starting 1937, both population and agricultural censuses give data on agricultural population. The data for 1960-84 are taken from FAO *Production Yearbook*, several issues, and FAO *Country Tables* 1987. Reprinted with permission, El-Ghonemy, 1992, *Journal of Agricultural Economics*, Vol. 43, No. 2

Having been marginalized by the State in the sale of its reclaimed land at concessional terms[5], the numerous small tenants and wage-dependent landless workers, who had no other activities, had virtually zero opportunity to buy land. The average price of land purchase, in 1947-1952, was equivalent to 20 years' average annual rent in real terms (El-Ghonemy, 1992, Table 4). Likewise, an adult male laborer (daily wage LE 0.11 during the same period, and assuming 210 working days *per annum*) had to accumulate all his wage earnings, without spending anything on living costs, for 22 years to purchase one feddan (average price LE 515). If he spent half his wage earnings, and assuming a constant wage-land price ratio, the period would double. In the face of the large landowners' monopoly power in their localities (having the right to evict their tenants at any time without payment of compensation, and to charge exorbitant cash rental), most tenants were unable to accumulate surplus from net revenue after rental payment. Nor could tenants and landless workers purchase land against borrowing from formal credit institutions which required landownership as collateral.

Considering that 60.7 percent of total cultivated area of Egypt in 1949/1950 was leased out by absentee landowners under insecure tenancy arrangements, and that about 45 percent of total rural households were landless (El-Ghonemy, 1953), the pre-1952 agrarian system was detrimental to rural development. It illustrates John Stuart Mill's statement that the State tends

> to augment the incomes of landlords; to give them both a greater amount and a greater proportion of the wealth of the community, independently of any trouble or outlay incurred by themselves. They grew richer, as it were in their sleep, without working, risking, or economising. (1886, Book V, p. 160)

Post-1952 Reform

Pro-*fellaheen* equity in rural development was among the immediate objectives of Nasser's revolution of July 1952. For the first time in Egyptian history, a maximum limit on private landed property was established by the land reform Law of September 1952. The ceiling was initially fixed at 200 feddans (84 hectares) per household head, lowered in 1961 to 100 feddans. Furthermore, the royal family estates were confiscated and ownership of land by foreigners was prohibited. *Waqf* land amounting to 11 percent of total cultivated area was acquired by the State. Tenancy arrangements were regulated, and the security of tenure for a renewal period of 3 years was provided. Rental values of leased land were considerably reduced and fixed at seven times the assessed land tax and 50 percent of harvest in the case of sharecropping. Private property in land, though restricted in size, was maintained and established in the Egyptian constitution of 1956.

These reforms were accompanied by two sets of public action. One was an accelerated reclamation of state-owned land, using the increased Nile water supply resulting from the construction of the High Dam in the 1960s. The other was the control of production and marketing of the main crops whose prices and areas are determined by government. At the village level, subsidized means of production are provided to farmers who are required to join government-patronized cooperatives. These local institutions are assigned a wide range of tasks: the registration of leasing contracts; the allocation of land among crops; the provision of basic inputs; and the procurement, at an administered price, of certain portions of the major food crops and all cotton and sugarcane harvests.[6]

By 1985, land reform laws had been completely implemented. Accordingly, 864,521 feddans or 14 percent of total cultivable area was redistributed to former tenants in units of 2-3 feddans per household (about one hectare on average). As in Morocco and Tunisia, the State retained and managed directly 12 percent of the expropriated lands, mostly orchards. The beneficiaries numbering 346,469 represented merely one-tenth of total agricultural households. In addition, out of a total state-owned reclaimed area of 1.4 million feddans, only 15 percent or about 215,000 feddans were allotted by 1988 in individual units of 4 feddans on average to some 54,000 households (*Statistical Yearbook*, 1989, Tables 2-22 and 2-29). Yet, nearly one-third of agricultural households remained tenants, and another 35 percent landless wage-workers. At the same time, the State has remained the largest single owner of reclaimable lands, manages a number of large farms and retains part of the expropriated lands.

Since the mid-1970s, the State open door policy (*Infitah*) has encouraged investment and trade by the private sector. With mounting economic crisis, this policy has, since 1986, been re-inforced by the government's gradual acceptance of the World Bank-and IMF-induced economic adjustment program. Accordingly, the government control of complementary production inputs has had to be reduced, and their subsidization removed. Likewise, crop prices (except cotton and sugarcane) are to be determined by market forces. The adjustment program includes ending the government-monopoly in agricultural credit supply and in land reclamation. It includes also the sale of state-owned farms in public auction. Simultaneously, the lobby of landowners and the syndicate of agricultural graduates has gained strength in influencing land policy. Landowners are demanding the right to evict tenants and the determination of rental values by the lease-market mechanism, and graduates want to own a part of state-owned land.[7]

Food Production Consequences of Land Policy

Because of the long span covered in the preceding review, we consider four periods: 1810-1835; 1835-1897; 1897-1952 and the post-1952 period. This breakdown is based on stages of development and government intervention. Let us briefly take them in order.

Pre-1952 Situation

During the first period (1810-1835), state monopoly in marketing and distribution of land rights was absolute, and production relations were feudal. Government bureaucrats and village *shaikhs* (headmen) used the corvée system of compulsory agricultural labor, which, like all forms of slavery, is prohibited by Islam. In addition, the farmers cultivated land under government tight directives. In such a situation, disincentive and resignation prevailed among the *fellaheen*.

During the second period (1835-1897), the government's control over the rural economy gradually gave way to market forces. Substantial public investment in irrigation and the inflow of foreign capital led to a fundamental transformation of the wheat-based subsistence agriculture into a dual economy having cotton as the engine for export-led growth, linked with the world market. The area growing cereals nearly tripled, and the areas of sugarcane and cotton increased 4-fold. Considering the weakness of statistics in the first and second periods, the broad trends in productivity per land unit (embodying capital investment) and *per capita* suggest that Egypt's food production and agricultural development proceeded at a pace unprecedented in the country's modern history. Agricultural output per head of rural population rose nearly six times, and land productivity per feddan rose sevenfold (1821-1878), according to O'Brien (1968, pp. 184-186).

During the third period (1897-1952), particularly after the First World War, all indices of agricultural growth slowed down, despite the use of chemical fertilizers and continuous land reclamation (Hansen and Marzouk, 1965, Chart 3.1 and O'Brien, 1968, p. 189). In the meantime, rural population almost doubled and total food output increased by only 30 percent. The diminishing food production *per capita* total population and productivity of agricultural labor were examined by El-Imam (1962), Hansen (1966) and Mead (1967), through different statistical techniques. The average, daily wage index in real terms for 1937-1943 was at its 1920 level and the corresponding marginal productivity of labor index was 76.5 percent of its base year level in 1914 (Hansen, 1966, Tables I and V).

Thus a Malthusian situation existed with a growing poverty among the *fellaheen* whose physical productive capacity was low due to a high incidence of ill health, undernutrition and illiteracy (El-Ghonemy 1953, pp. 70-72).

Elsewhere, the author estimated the proportion of rural population living in absolute poverty in 1950 at 56.1 percent. This was based on a poverty line of a minimum annual income of 35 Egyptian pounds per household established in 1949/1950 by Egypt's Social Security Scheme (*al-Damān al-Igtimā'ee*). The poor were landless workers (63%), sharecroppers (2%), owners of less than 0.4 ha. and pure tenants holding less than 0.7 ha. (24%), nomads (2%) and disabled heads of rural households (4%) (El-Ghonemy, 1990, p. 247).

Has this high incidence of rural poverty been reduced, and food production increased by the post-1952 series of agrarian reforms and the government's tight control of the rural economy?

Post-1952 Food Production

In a country already densely populated, annual rates of population growth accelerated from 1.3 percent in the first half of this century to 2.2 in the 1960s and reached 2.8 percent between the two population censuses of 1976 and 1986. In 1950-1988, the rate of change of productivity per working person in agriculture fell from 2.0 percent per year to 1.2 percent. Likewise, aggregate food production failed to keep pace with population growth and food productivity *per capita* total population fell sharply from the annual rate of 0.6 percent in 1961-1970 to a negative 0.7 percent in 1980-1984 (FAO Index 1974-76 = 100).

Why has food production fallen? Firstly, distorted price policy has led to a decline in the area of wheat and beans (*foul*), the two major staple food crops. Cotton, the major export crop fell also by 43 percent, while the area of green fodder (*barseem*) rose by 113 percent (livestock products are excluded from price control). These changes are shown below.[8]

	Average in Thousand Feddans		
	1947-1952	1982-1986	Change %
Cotton	1,806	1,036	-43.0
Wheat	1,447	1,253	-13.4
Broad beans (*foul*)	356	318	-11.0
Barseem Mostadeem (green fodder)	906	1,926	+112.6

In response to the government pricing policy (procuring harvests at prices 50-70 percent below market levels), the enterprizing farmers have tended to grow more of the profitable *barseem* and less of wheat and beans. Consequently, the rate of self-sufficiency for wheat fell from 70 percent in 1960 to 24 percent in 1980 (Wally, 1982, Table 1) and self-sufficiency rate of all cereals, including rice, fell from 77 percent in 1969-1971 to 49 percent in 1983-1985. This has led to widening the gap between cereal production and total requirements, and to increasing the reliance on food imports and the politically vulnerable food aid (see Chapter 3, Tables 3.2 and 3.3).

Secondly, soil fertility of over-irrigated land has declined due to delay in drainage schemes, leading to soil salinity and a fall in yields. With the reduction in cereal areas and yields, the annual rate of growth of total volume of cereal harvests fell sharply from 3.3 percent in 1960-1970 to 0.9 percent in 1981-1988 (FAO, *1990 Country Tables*).

Distributional Impacts

Unfortunately, no reliable data exist on households' personal income distribution in agriculture during the pre-1952 period. The incomplete estimates of the Ministry of Agriculture (1945-1947), which did not include hired laborers' income, suggest a Gini coefficient of 0.65 (El-Ghonemy, 1953, pp. 120-123). In this state of an agrarian economy, income distribution reflected the concentration of the size distribution of landownership, which ranged between 0.6 and 0.7 in terms of the Gini coefficient. This is not surprizing in the face of the functioning of the land market and the state of polarization of agrarian structure described earlier.

TABLE 6.2 Rental Value, Daily Wage Rates, Cotton Price, Gross Output Value and Density Ratio in Egypt, 1913-1986

Year	*Average Nominal Values*					
	Rental Value LE/year	*Adult Male Wage LE/day*	*Output Value LE (1)*	*Cotton Price LE (2)*	*Rural Population Density (3)*	*Deflator (4)*
1913	9.0	0.040	10.3	3.6	0.60	100
1927	7.1	0.042	9.7	5.9	0.53	98
1929	9.0	0.050	9.9	4.1	0.53	94
1933	5.1	0.022	10.2	2.5	0.51	58
1937	5.7	0.030	9.1	3.2	0.46	86
1939	7.2	0.035	11.2	3.2	0.45	108
1943	15.2	0.063	18.6	6.9	0.45	205
1944	18.0	0.093	22.7	7.9	0.45	225
1945	19.0	0.093	26.4	7.5	0.45	225
1946	19.4	0.095	27.0	8.4	0.44	256
1947	22.0	0.095	28.2	12.3	0.46	241
1948	23.3	0.100	29.5	11.1	0.45	234
1951	34.0	0.126	37.4	22.1	0.43	227

(continues)

TABLE 6.2 *(continued)*

Average Nominal Values

Year	Rental Value LE/year	Adult Male Wage LE/day	Output Value LE (1)	Cotton Price LE (2)	Rural Population Density (3)	Deflator (4)
1952	33.0	0.120	34.5	12.3	0.43	228
1955	25.0	-	33.6	14.2	0.41	253
1956	26.0	0.100	39.7	17.0	0.39	294
1960	25.0	0.125	43.8	15.0	0.37	290
1965	22.5	0.220	57.3	14.9	0.34	447
1967	21.9	0.245	61.8	16.0	0.34	413
1970	26.0	0.250	70.0	18.2	0.34	496
1975	30.0	0.460	125.5	25.4	0.30	690
1980	52.5	1.370	271.4	47.2	0.28	1270
1981	53.6	1.810	307.8	58.1	0.27	1399
1982	63.7	2.350	373.2	60.0	0.26	1592
1983	66.0	3.090	434.8	65.1	0.25	2064
1984	69.0	3.830	485.0	74.0	0.24	2382
1986	80.0	4.700	535.0	97.2	0.21	3062

Notes: LE is the Egyptian pound, which equalled US$4.13 up to Sept. 1949, and was devalued to US$2.87 until 1977. Between 1982 and 1986, it equalled $1.22. (1) Gross value agricultural output (field crops, vegetables and fruits) per feddan cropped area. (2) The average price of one *Kentar* weight of cotton of different staple lengths. (3) Density ratio refers to cultivable land *per capita* rural population, the latter from *Statistical Yearbook*, several years. (4) The deflator is the cost of living index for rural areas established by the Institute of National Planning, spliced with 1966/67 = 100.

Sources: Rental values for 1913-29 are from Richards (1982, Tables 3.17 and 4.4) and El-Ghonemy (1953, Tables 10 and 37 for 1929-51). The rest are from the Bulletin of Agricultural Economics, several issues, Ministry of Agriculture (in Arabic). Adult male wages for 1912/13 and 1927-1930 are from Hansen (1966, Table III), 1937-1951 are average rates collected by the author from 98 villages in lower Egypt and 83 in Upper Egypt during his work in the Fellah Department. The rest are from Radwan (1977, Table 3.2) and the Ministry of Agriculture. Gross value of output for 1913 and 1927-1930 is calcualted from El-Imam (1962), to which we added 12 percent, the gross value of *Barseem* which he excluded; and for 1937-39 is based on Anis estimates, cited in Hansen (1966, p. 402). The rest are from published statistics of the Ministry of Agriculture. Cotton price is from Mead (1967, Table V.A6) for 1913-1956, and the rest from the Ministry of Agriculture. The deflator is from *Statistical Yearbook*, CAPMAS, Cairo. Reprinted by permission, El-Ghonemy (1992), *Journal of Agricultural Economics*, Vol. 43, No. 2, pp. 175-90.

Income Transfers (1952-1986)

Land reforms of 1952 and 1961 transferred landed property from rich landowners to about one third of total tenants in family farm units of one hectare on average. The rest of the tenants were protected -- paying reduced rent, and they have virtually become irremovable. The affected rich farmers owning land above 100 feddans, who were proportionately small (0.1 percent of total owners), were hit hard as the compensation paid in unindexed government bonds was minimal, and the reduction in rental values below their market level was considerable in real terms. The State sold the requisitioned land to the beneficiaries at less than half the average market value, payable over 40 years by annual instalments, bearing a subsidized interest at half the market rate. The sale price was successively reduced by the government to 25 percent of its original level, and the new owners were also exempted, until 1986/1987, from land tax payment.

New Owners' Gain. The gain is represented by the difference between the annual instalment and the previously paid rent plus expected increase in yields minus taxes (land, defence and municipal taxes). Our sample survey of 611 household beneficiaries conducted in 1973 (20 years after the implementation of land reform) found that the new owners' average *per capita* real gross income was 189 percent higher than in 1953, suggesting an average annual increase of 4.1 percent.[9] Given the different deflators used, this rate was higher than the 2.3 percent average annual growth of agricultural GDP (1953-1975) calculated by the Ministry of Planning at 1959/1960 fixed prices.

Our sample survey also indicates that non-land assets (livestock heads per household) and non-agricultural income increased substantially, compensating for the smallness of distributed units of land (El-Ghonemy, 1990, pp 234-236). It was also found that new owners gained from yield-increasing technology provided by the Agrarian Reform Authority. Cotton and rice yields were higher in the sampled areas than the national average. Yields of wheat, maize, beans and sugarcane were around the national mean. Other micro-economic studies conducted by Egyptian scholars in the 1950s and the 1960s also confirmed that the new owners gained from increased yields.[10] However, it cannot be asserted that productivity gains were universal in the land reform sector. Nor can the incremental output brought forth by land reclamation and technical change be separated from land reform effects.

Tenants' Gain from Rent Control. Sustained gains from the government-administered land-lease market are substantial, considering the large number of tenants (about one million or 62 percent of total landholders in the 1961 agricultural census). Thus, the gains from rent control are of a greater scale than those from the limited land redistribution program. To estimate the tenants' gain in real terms, two methods are employed. One is nutritionally based and expressed in maize equivalent of rental values before and after rent control. This estimate is calculated from maize prices per *ardab* given in *Annuaire Statistique* and the *Agricultural Economics Annual Bulletin*. Maize is a staple food in rural areas, and *ardab* is a unit of capacity and equals 140 Kilograms weight. The changes before rent control (average 1949-1952) and after (average 1983-1986) are as follows (the deflator is in brackets and rental values are taken from Table 6.2).

	1949-1952	1983-1986	Reduction % 2 over 1
	(1)	(2)	
Rent, average nominal value LE	32.5	72.0	
Deflated by cost of			
living index 1913 = 100	14.4	2.8	80.5
	(226)	(2504)	
Measured in Maize/Kg. equivalent	1739.0	446.0	74.0

The above set of data shows that, whereas the nominal (current) value of annual rent per feddan more than doubled between the two periods, it fell substantially in real terms. This result is confirmed by the other method for estimating tenants' gain, in which we used the moving average growth rates of rental values and indices of inflation, together with the geometric mean of the latter. The data used and the results obtained are given in Table 6.3. Clearly, there is an overlapping in the indexes used in both methods; maize price is a main component of rural consumers' cost of living index, and is reflected in crop prices index.

Depending upon the accuracy of official statistics on average current values of rent, the considerable reduction of deflated rent estimated by both methods represents an outright transfer of *real* income from absentee landowners to tenants. In most cases, the fixed rent seems to be enforced, particularly in areas administered by co-operatives, government agencies and public agricultural corporations. In some localities, however, a 'black market' has emerged, and rent is illegally higher. Looking closely at the data given in Table 6.3, we find that, except in 1975-1980, following the introduction of a trade liberalization policy (*infitah*) by the Sadat regime, rent grew at rates lower than those of all indexes of inflation. In real terms, rent reduction ranged from 15.4 percent *per annum* in 1960-1965 to 7 percent in 1980-1986. The considerable transfer of income from landowners to tenants is also manifested in the sharp fall in the share of rent in output value per feddan from 70-80 percent before the 1952 land reform to only 15-18 percent in the 1980s (see Table 6.2).

TABLE 6.3 Tenants' Gain in Real Terms from Rent Control in Egypt, 1955-1986

Sub-periods	Rent Nominal Value (1)	Cost of Living Index (2)	Wholesale Prices Index (3)	Crop Prices Index (4)	Inflation Rate (5)	Gain % (6)
	Moving average annual growth rates (percent)					
1955-1960	1.0	8.0	2.4	n.a.	n.a.	3.4
1960-1965	-10.0	10.5	2.8	n.a.	n.a.	15.4
1965-1970	1.7	1.0	3.5	5.3	3.5	1.1
1970-1975	1.6	6.2	6.8	11.0	5.2	5.4
1975-1980	13.6	13.5	12.9	12.6	10.0	-1.4
1980-1984	7.2	15.2	10.4	13.3	12.4	5.5
1984-1986	8.0	17.2	13.8	21.0	15.0	8.5

Notes: n.a. stands for not available. The gain in column 6 is a subtraction of value in column 1 from the geometric mean of the rates of the deflators in Col. 2, 3, 4 and 5.

Source: M.R. El-Ghonemy, "The Egyptian State and Agricultural Land Market: 1810-1986", *Journal of Agricultural Economics*, Vol. 43, No. 2, 1992. Reprinted by permission.

Distorted Economic Relationship Between Land Productivity and Rent

In spite of its substantial distributional benefits to tenants, rent control has distorted the relationship between rental value and the profitability of land in terms of the value of output per feddan of cropped area.

Since the work of the classical economists, total agricultural output has been recognized as the source of factor shares (labor wages, rent and profits). In a private-landed property economy, one would expect that value of output and rent per feddan are positively correlated, and that the variation in rent is mostly determined by changes in output value. One would also expect in the Egyptian setting of persistent pressure on the scarce cultivable land, that the lower the ratio of land *per capita* rural population is, the higher the rent would be.

For understanding the post-1952 changes introduced by rent control, the empirical situation which existed between 1913 and 1986, is graphically presented in Figure 6.1. Nominal values (at current prices) of rent, output per feddan and cotton price given in Table 6.2 are deflated by the index of rural consumer cost of living (1913 = 100) and plotted in the Figure. It shows the expected positive relationship between rent and the other two variables, when market forces were dominant (1913-1952). After the extensive government intervetion, output value of land (but not cotton price) has lost its important economic role, and contrary to expectation, the rent-output positive relationship is reversed. In estimating these relationships, technical change (irrigation, drainage, improved seeds, fertilizers, etc.) is assumed to be captured by variation in both output and density on cropped land. The author has, elsewhere, statistically estimated these relationships, and explained their important implications for resource use efficiency, incentive structure and income distribution in rural Egypt.[12]

Table 6.4   Changes in the Distribution of Land and Income, Landlessness and Poverty Levels in Rural Egypt, 1950-1982

Measurement	1950	1965	1982
Inequality index			
Landownership distribution	0.740	0.384	0.432[a]
Rural income/expenditure			
distribution	0.650[b]	0.290	0.340
Poverty Levels			
Rural pop. (thousands)	13,710	17,754	23,760
Percentage of rural poor	56.1	23.8	17.8
			24.2[c]
Rural poor (thousands)	7,691	4,225	4,229
			5,100[c]
Landlessness			
Landless households as %			
of total agric. households	59	40	27

Notes: a. refers to 1984. b. refers to 1947. c. Another estimate made by the World Bank. For an explanation of this divergence, see Chapter 7, p.136.
Inequality index is the Gini coefficient.

Sources: The Gini coefficient of inequality of income distribution for 1947 and land distribution for 1951, 1965 and 1984 is calculated by the author from data compiled by the Ministry of Agriculture (El-Ghonemy, 1953), and *Statistical Yearbook*, CAPMAS, several issues, respectively. Using rural household expenditure surveys conducted by the Central Office of Statistics (CAPMAS), Richard Adams, (1985, Table 4), calculated the Gini index for the other years. Poverty levels - 1950 are estimated by the author, see El-Ghonemy, 1990, p. 247. For 1965 and 1982, see Adams, 1985, Table 1. The 1982 higher estimate of poverty made by the World Bank is taken from its *Country Study, 1991,* Table 2.1. Landless households - 1950 and 1965 are from Radwan (1969). The estimate for the year 1982 is calculated by the author from the 1982 census of agriculture and the data on agricultural population in *Country Tables,* 1990, FAO. The number of landholders is substracted from total agricultural households, assuming the size of family to be 5.3 persons (*Statistical Yearbook*, CAPMAS, 1989, Table 1-7).

Rising Inequality and Poverty

Despite their limitations, data given in Table 6.4 suggest that inequality of distribution, which was sharply reduced between 1952 and 1965, increased between 1965 and 1984. The Table also shows a substantial reduction in the incidence of rural poverty between 1950 and 1982. Although the several estimates of poverty are based on varying assumptions on average size of rural/agricultural household and minimum income/ expenditure (poverty line), they provide information for judging the approximate order of magnitude of change in proportionate terms and absolute numbers.

Why, then, has inequality of both land ownership and income distribution in rural Egypt increased between 1965 and 1984? We begin with offering a few explanations. They are in conjunction with changes in the land market structure combined with agricultural policy which influence the allocation of land among crops.

The reversal shifts in private land transactions has taken place since the mid-1970s, when land speculators' profits have attracted some peasants to sell parcels of their land at inflated prices. These transactions increased the share of middle size landowners (5-20 feddans) in total area from 17.7 percent to 21.3

FIGURE 6.1 The Movement of Deflated Nominal Values of Rent, Gross Output per Feddan and Cotton Prices in Egypt, 1913-1986

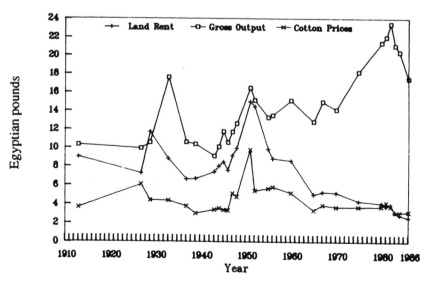

Source: Data given in Table 6.2. This graph is taken from El-Ghonemy, 'The Egyptian State and Agricultural Land Market, 1810-1986, *Journal of Agricultural Economics*, Vol. 43, No. 2. Reprinted by permission.

percent, and reduced the area owned by small farmers (below 5 feddans) by 268,000 feddans between 1961 and 1985 (*Statistical Yearbook*, 1989). A further possible explanation of rising inequality lies in non-market land transfer, i.e. the shift -- in state policy concerning the distribution of newly reclaimed land -- away from according landless farmers top priority (established in late 1950s) towards its sale to the highest bidders in open auction.

Moreover, crop pricing-cum-land allocation policy has led to a heavy indirect taxation of producers. Small landholders (cultivating traditional crops whose price is controlled by government) were hit more than larger landholders growing relatively less of these crops and more of the non-controlled crops (livestock feed, vegetables and fruits). Ibrahim (1982, Tables 2 and 3) found that indirect taxation -- ranging from 22 to 84 percent of farmers' gross income from the price-controlled cotton, rice, onions, beans and sugarcane -- has widened inequality among farmers by their size of holdings and crop mix.

With regard to poverty reduction between 1950 and 1982, there are interconnected possible determinants. Apart from redistributive benefits accrued from land reforms, there has been a substantial change in the rural labor market leading to a steady rise in real wages, which tripled between 1952 and 1986 (see Table 6.2). Likewise, remittance receipts by migrant rural workers earned in the oil-rich Arab States have significantly increased their households' incomes (Adams, 1991, Table 7). Two further determinants of poverty reduction are important. One is the expanding non-land asset ownership by low-income farmers, notably livestock and tractors for hiring out (Commander, 1987, Tables 9.5 and 9.12). The other is the total GDP growth at an annual rate of 8.5 percent in real terms between 1973 and 1984. This rapid growth has enabled the government to continue its generous food subsidies and free health and education services, enhancing the *fellaheen's* abilities. Free, compulsory public schooling has expanded access to renumerative employment-opportunities outside agriculture, of which the children of the rural poor availed themselves.

The Prospects for The Rural Poor

Past and current events in the Middle East suggest the volatile and uncertain nature of remittance receipts from international labor market which can no longer serve as an adequate and secure safety net to resolve Egypt's poverty problems. The sharp fall in oil-export revenues of Arab States, resulting from the collapse of World oil-pricing system in the mid-1980s has reduced the demand for workers from Egypt by about 30 percent in 1986.[11] Most of the workers returned from Iraq in 1989 and 1990 after its war with Iran ended, and from Kuwait and Iraq after the start of the Gulf crisis in August 1990. The resulting collapse of the labor market in the Gulf has aggravated the would-be migrants' deprivation. It is also hardly possible to forget the sharp fluctuation of the number of Egyptian workers in Libya in response to the frequently changing political relations.

Confronted in 1986 with a deep recession manifested in a budget deficit reaching 23 percent of GDP, external debts amounting to US$ 30 billion and annual inflation of 13-15 percent, the government found it necessary to gradually accept the IMF and the World Bank's macro-economic adjustment program and its conditions. The acceptance of the program between 1986 and 1991 is likely to end most of the subsidized social development benefits such as cheap food, social services, and means of agricultural production. Substantial cuts in public expenditures for land reclamation and new settlement schemes are also included in the program.

Egypt is a land-scarce economy afflicted by a rapid population growth, increasing scarcity of water for irrigation and steadily deteriorating annual rates of economic growth from 9 percent in 1974-1981 to 5 percent in 1982-1986, and declined further to 1.0 percent in 1989. In the meantime, the country's dependence on food aid has considerably increased. Moreover, falling migrant workers' remittances from 13 percent of GDP in 1979-1984 to 11 percent in 1990 (and to a staggering low of about 7 percent in 1991, because of the Gulf War), have adverse, widespread effects on the economy and the welfare of rural people. With such a state of the economy at the start of the 1990s, coupled with the revival of the large landowners' lobbying strength, rising inequality, malnutrition, poverty in rural areas are more likely to persist in the 1990's.[13] In fact, the World Bank in its recent study on the structural adjustment program in Egypt expects rising inequality and worsening poverty conditions. It admits that the recently introduced program "will hurt people who are already poor and are at risk of becoming poor" (1991, p. 93).

The present recession and the expected adverse effects of economic reforms pose a challenge to the country leadership, and present a dilemma over hasty economic reforms needed to improve the performance of the economy, and at the same time to sustain the post-1952 state commitment to realize social justice and tangible evidence of a speedy poverty alleviation. This commitment is stated in the Egyptian constitution and is recently restated as an objective to be realized in the Five-Year Plan, 1988-1992: "to make social services available to everyone and to improve income distribution". But to improve the adverse distributional effects of the recently implemented economic reforms, a long gestation is required. For socio-political stability purposes, the country cannot afford this long-term solution. At the time of writing (December 1991), there is a mounting domestic pressure for public action to alleviate hardships resulting from rising costs of living, unemployment and poverty. Non-governmental organizations and the Moslem fundamentalist movement are expressing, in different ways, discontent and anxiety over the juxtaposition of affluence and poverty in an underdeveloped economy, whose GNP *per capita* in real terms has declined in the 1980s, and in which 75 percent of the total number of the poor live in rural areas.

In particular, it would be difficult, through the market mechanism favored by the recent structural changes, to sustain the tenants' gains, and to satisfy the *fellaheen's* persistent demand for the scarce cultivable land. This paradoxical situation has been compounded by the fall in land supply brought about by both a declining availability of water for irrigation and the absurd public and private transformation of scarce fertile land into non-agricultural use at the annual rate of nearly 30,000 feddans, including the removal of top-soil for making bricks (Biswas, 1991, p. 19). This dismantling of a social asset, created over many centuries from the Nile water silt, has occurred while 95 percent of Egypt's total land area is desert. In sum, persisting population pressure on the declining effective supply of land, the falling food productivity, the slow industrial growth and the collapse of the international labor market suggest that, in the 1990s, the prospects for the rural poor and landless farmers are gloomy.

Sudan

Development literature has tended to characterize the problems of Sudan in terms of frequent outbreak of famine, increasing concentration of profits and scarce capital in a few hands, civil war, and political instability. Likewise, media coverage, television and vivid reporting have brought all kinds of Sudan's human sufferings right into the homes of millions in the world. Being one of the world's least developed countries and with an overwhelmingly rural population, these development problems seem to be rooted in Sudan's social structure and rural underdevelopment. This section attempts to explore these issues.

Out of the North African countries, Sudan is endowed with vast amounts of water resources and potentially cultivable land. Its livestock wealth is the largest in Africa and the numbers of cattle and sheep are nearly four times as much as the total number in the other five North African countries. Yet Sudan cannot feed its own people, without food aid and concessional imports of wheat and dairy products. Sudan is also the only country in North Africa whose leadership did not give prominence to land tenure issues at the time of independence in January 1956. Since then, there has been a lively debate and conflicting views among Sudanese scholars on how to restructure the agrarian system for alleviating food insecurity and persisting rural poverty.[14]

Roots of Dual Rural Economy

The performance of Sudan's rural economy reflects its dual composition of two distinct sectors. In broad terms, they are the modern or technology-based sector of irrigated and mechanized rain-fed farms on the one hand, and the traditional sector, producing food crops and raising livestock with a minimal technological change, on the other. Only 10 percent of total rural population live in the former and the rest, including small farmers and nomadic pastoralists, live in the traditional sector.

This dual structure has been rooted in historical forces and in changing state policy with regard to priorities in agricultural development. Therefore, in order to understand the duality of the rural economy and its constituents (agricultural growth, income distribution, food production, land tenure, labor market, credit and marketing facilities, poverty, etc.), we need to comprehend the sequential events of policy-making that have shaped the present conditions of rural underdevelopment.

The Land Tenure Dimension

The historical experience suggests that the present land tenure dimension of duality originated in the seventh century by the Arabs who immigrated from Egypt.[15] The small proportion of individual land ownership was brought about through the purchase of land from, and intermarriage with, the indigenous Nubians. During their movement to rain-fed areas in the Eastern and Western regions, the immigrants practised the communal land tenure system followed by their ancestors in the Arabian peninsula for grazing and growing cereals. Hence, the Islamic principles of inheritance, payment of tithe (*Ushr* land tax) and *Zakāt* were introduced. Throughout the Ottoman rule, starting in the sixteenth century, this tenure system continued, with the introduction of two additional institutional arrangements. One was the transfer of unregistered property rights in land to the State (*miri*). The other was the granting of agricultural land (*wathiqah*) by the Sovereign to Moslem leaders, influential heads of tribes, senior government officials and military men.

It was during the British administration[16] (1889-1956), that the salient features of the current duality of modern and traditional sectors were laid down. The colonial administration accorded priority to:

1. Payment of officially assessed taxes, and not the customary tributes.
2. The legal recognition of tribal rights in communally held land combined with the territorial definition of pastureland of each tribe.
3. Checking land sale from small owners to land speculators, who were motivated by the prospective high return on purchased land that would soon be irrigated by investing public funds and foreign private capital.

The speculators' action was also induced by high expectation of future gains from growing cotton, particualrly in the Gezira plain lying between the Blue Nile and the White Nile.

4. The transfer of unregistered lands into government property by the Land Settlement Ordinance and the Land Acquisition Acts of 1925 and 1930, respectively.

5. The creation of a modern sector of export-orientated irrigated cotton and the commercial, mechanized farming in rain-fed lands.

Towards State-led Capitalist Agriculture

The establishment of the Gezira Scheme south-east of Khartoum was a significant feature of the pre-independence agricultural development, through irrigation investment.[17] With abundant labor and fertile land suitable for growing cotton, capital and technical knowledge were lacking for employing the plentiful Nile water and linking agriculture with world market. The scarce capital was supplied by private investors from U.S.A., the British Cotton Growing Association and a loan from the British Government. The institutional structure was provided by: a concessional grant of 10,000 feddans of state-owned land in Zeidab near the Atbara river to serve as a pilot area; the formation of the Sudan Plantation Syndicate for the management of the joint venture; and renting irrigated land to local cultivators who also pay an established rate for pumped water.

Despite the British administration's corroborated fear of the transfer of the natives' land to foreign investors, the Gezira Scheme was gradually extended from 15,000 feddans in 1912 to 100,000 feddans in the early 1920s and further to 300,000 feddans in the 1930s. Before independence in 1956, its area reached nearly half a million feddans, including 230,000 feddans cotton. The acquired land was partly rented from native landowners, and partly expropriated against payment of full compensation. Irrigated land was allotted to natives in large units of 40 feddans (16 hectares) under long-term lease stipulated in the Gezira Land Ordinance of 1927. A rotation of cotton and food crops (beans and sorghum) was strictly administered. Half the cotton harvest was left to the tenants who conducted all field operations under intensive supervision by the Gezira administration, leaving a minimal managerial responsibility to the individual tenant.[18]

The private pump-scheme was the second technological change for expanding irrigated cotton. This fast growing enterprise had considerably expanded the capital market, and the area growing long-staple cotton in the Blue Nile province increased 32-fold from 3,591 feddans in 1932 to 113,883 feddans at the time of independence. Likewise, the scheme induced new institutional arrangements for licensing imported pumps, borrowing credit linked to cotton-marketing and for holding irrigated land under tenancy (Osman, 1987,

Table 1). Big Sudanese cotton growers and merchants as well as foreign investors accumulated profits from these transactions.

The third form of capitalist development was the introduction in 1944/1945 of mechanized farming in rain-fed areas of Gadaref district (between Gezira and the Red Sea)for the tractor-based production of sorghum and sesame. Starting with leased units of 30 feddans, the tenancy area expanded dramatically to nearly one thousand feddans at a very low annual rent of 3 piasters per feddan (One Sudanese pound £S = 100 piasters). Simpson reported that the tenants were mostly large landowners and merchants having capital and managerial ability. He says "Some of these early pioneers had over 20 tenancies each, and had encroached on undemarkated land, including the corridors left for the passage of nomads and their animals. They cleared trees and bushes left as shelter belts for nomads and their cattle" (1987: pp. 274-5).

Post-independence Perpetuation of Duality

Since 1956, when Sudan gained independence, the above broad features of rural development policy, regarding land tenure system and the pattern of agricultural growth have virtually continued. However, there has been a series of swings in ideological preference behind the pace and content of state policy. The shift in public action has resulted from abrupt changes in the form of government (authoritarian, military and parliamentary). The role of the State in agriculture, in general, and in South Sudan, in particular, has also been influenced by the frequent swings in the country leadership's ideology (Sudanese socialism, social ownership, private sector-based capitalist economy, and Islamic fundamentalism). For example, the Islamic Charter Front after joining the government has, since 1981, influenced private investment and rural credit market, and some people say has contributed to the outbreak of civil war in Southern Sudan, to which Islamic laws are alien. With these factors in mind, the main elements of the post-independence strategy for rural development are briefly presented.

The Expansion of the Modern Irrigated Sector

The Sudanization of the Gezira scheme (GS hereafter) management was a priority in post-independence policy. Its technical and institutional bases were extended into the domestically financed Managil district scheme covering an irrigated area of 80,000 feddans. Between 1957 and 1962, the latter allotted to each tenant 15 feddans (6.2 hectares) compared to 40 feddans in the original GS. By reducing the unit size, the government was able to absorb more tenants in the scheme, and reduce their reliance on hired labor. The GS triple-partnership model based on risk-sharing (Government, Gezira board and the tenants) has been maintained. Whereas the tenants' share in cotton proceeds has increased from 40% to 44% and their participation in decision-making improved, the centralized control of inputs' supply and marketing of harvests remained virtually unchanged. An interesting difference between the Gezira and Managil schemes is the higher proportion of female tenants in the latter; 10.2 percent compared to 6.9 percent in the Gezira.[19] They are, however, underrepresented as working women represent about 26 percent of total labor force in agriculture.

Irrigation-pump schemes also have been substantially enlarged along the Blue and White Nile in private and state farms. The former (Blue Nile) irrigating about 350,000 feddans followed the previously established arrangements for licensing and distribution of responsibilities between pumps' owners and their tenants for growing cotton and food crops (sorghum, groundnuts and vegetables). Ownership and management of small-pump schemes were encouraged by the government, through organizing co-operative societies. Larger ones (exceeding 6 inch diameters of pump) were nationalized by the Nimeiri regime against payment of compensation. The tenants remained the labor suppliers, having a minimal managerial responsibilities, whereby the government corporation makes major production decisions. The government also monopolized the supply of technical inputs for cotton and its marketing (Magar, 1986, pp. 180-1).

Expanding Mechanized Farms to the Nomads' Disadvantage

Mechanized rain-fed schemes in the Savannah belt have rapidly expanded. The government's role has ranged from securing international inflow of capital and providing public expenditure for infrastructure, to arranging contractual transactions with lease-holders. By 1979, the area of these schemes was over three million feddans (1.25 million hectares), mostly in the Kassala province and Western provinces of Kordofan and Darfur. In the 1980s, the cultivated area expanded rapidly to nearly 8 million feddans and produced 57-60 percent of total harvested sorghum and millet. Likewise, expanded areas of edible oil crops (sesame and sunflower) produced nearly 17 percent of total oil seeds in 1987-1988.[20]

Part of the land acquired for mechanized farming came from a deliberate encroachment on the communally held tribal lands. Consequently, the nomads have been pushed to areas with lower rainfall, and some became hired workers in these schemes. Lease-holders, allotted 1,000-1,500 feddans each for growing sorghum and sesame, are mostly urban merchants, former government officials, members of the armed forces, and private professionals. They have little, if any, experience in agriculture and farm management. They took advantage of the attractive government facilities: subsidized credit; preferential access to capital equipment imported by scarce foreign exchange; and the provision of land already cleared of trees and complete with required infrastructure carried out by the state-controlled Mechanized Farming Corporation. Yet these generous subsidies were not matched by annual rental values paid by the tenants to the government, amounting to only £S 5 per feddan on average, or 40 cents at the commercial rate of exchange in 1990.

Increasing Involvement of Multinationals in Agriculture

Faced with the pressing need for foreign capital, the government induced multinational corporations through a wide range of concessions granted by the 1967 Act on investment. The concessions include free irrigation water, duty free imported equipments, repatriation of profit and imported capital, and a 5 to 10 year exemption from business profit tax. The several investment acts were later consolidated in a single Act of 1980 that also guarantees foreign investors against nationalization. The collusive multinationals' alliance with their Sudanese affiliates has made considerable monopoly profits from sugar plantations, large-scale production of sorghum and sesame, importation of tractors and sorghum seeds for the mechanized rain-fed schemes, and fattening of sheep and cattle.

These areas of concern suffer from a scarcity of hard evidence, particularly with respect to the net social cost and gains to the Sudanese rural economy. Apart from information on the secured generous concessions, there is only a few descriptive studies on the extent of the involvement of multinational corporations. For example, Tetzlaff (1984) reported that they held large plantations of sorghum and sesame covering an area of 250,000 feddans, and controlled the supply of imported tractors at the rate of 900 tractors per year, each costing US$ 10,000-16,000. Another example is that of one corporation, Lonrho, which was given in 1972/1973 the absolute monopoly power to import all capital goods, as well as to cultivate 82,000 feddans of sugarcane, and to process the harvest (Abdul-Karim, 1988, p. 42).

Establishing Settlement Schemes and Controlling Production

During the 1960s, a few planned schemes, including the large-scale Khashm El-Girba (150,000 ha.) and the small Undokoro scheme (9,400 ha.) were established to sedentarize certain groups of nomadic pastoralists.[21] The former scheme was implemented in consequence of the concluded 1959 Nile Water Agreement between Egypt and Sudan. Khashm El-Girba began in 1964 for the urgent resettlement of 10,000 farming families in Wadi Halfa district North of Sudan which was entirely inundated, following the construction of the Egyptian High Dam at Aswan. Egypt paid 70 percent of total cost, amounting to 23 million pounds Sterling (£S equalled almost one Sterling or US$ 2.87 in the 1960s). Each household was allotted 15 feddans (6 ha.) of irrigated land under lease arrangement, in addition to a private plot of 0.5 - 1.0 feddan for growing vegetables.

The rest of the land was also leased in 15 feddan units to nomadic people from the Shukriyya tribe. During the author's field visit in 1974, it was found that these settlers, having no past experience in growing cotton and groundnuts, were instructed by the scheme staff to allocate two-thirds of their holdings to these two crops. Furthermore, it was found that about one-third of the settlers (*halfawiyin*) were absentee, and some resident tenants were operating 5 holdings each worked by hired labor, which contradicted the scheme's aim. Despite the heavy capital expenditure in irrigation and current public expenditure on technical supervision, average yields of cotton, wheat and groundnuts were almost half those produced in the agricultural research station of the scheme. The settlers were also grouped into patronized co-operatives, cultivating their holdings under the strict control of three separate and poorly co-ordinated government departments, (irrigation, agriculture and local government), resulting in high administrative costs including accounting procedures, that was estimated at 26 percent of gross value output per feddan.

State Control of Production and Credit Supply

Apart from its strong control of cotton trade and management of large irrigation schemes, the State has controlled institutional agricultural credit supply and established large state farms in both irrigated and rain-fed areas. Specialized in the production of cotton and sugar, these farms have been expanded in numbers and size after the inclusion of the Sudanese Communist Party in government in 1964. An example of these state farms is the 50 thousand feddans Guneid farm growing cotton and sugarcane under irrigation. Instituted state farms in rain-fed areas such as Agadi, Habila, Simson and Guzrom are very large indeed. For example, the Agadi state farm covers an area of 105,000 ha. All are intensively mechanized, bureaucractically managed and over-staffed (Simpson, 1987).

Established in 1957, the Agricultural Bank (ABS) was given the mandate to favor small farmers and their agricultural co-operatives over large producers in mechanized farms and irrigated schemes. However, its operation has gradually shifted away from small farmers and towards financing larger farmers, influential sorghum producers in mechanized rain-fed schemes and traders, importing tractors and irrigation pumps. So did other commercial banks (Ali, 1980). Furthermore, in its grain market intervention, ABS supported primarily the few thousand sorghum producers at a high cost, reaching US $ 100 million in 1987 alone (Maxwell, 1991, p. 10 and p. 185). Moreover, the government's growing involvement in credit supply was reinforced by the establishment of the Public Agricultural Production Corporation for financing farmers in state-controlled schemes (e.g. Rahad, Gezira, El-Suki).

Adjustment of Lending Credit According to Islamic Principles. With the growing pressure of the National Islamic Front,[22] payment of interest on loans from ABS was prohibited by President Nimeiri's Decree of February 1981. It was replaced by a services charge, including premium and administrative costs. Following the establishment of the Faisal Islamic Bank in 1978 and the Islamic Investment Company later, some commercial banks adopted their procedures in savings, lending and in investment.[23]

However, amid these shifts in priorities and the ideology behind them, a large section of small farmers have continued to rely on the non-institutional credit arrangement (*sheil*). This long-established customary arrangement in rural Sudan is a crop marketing-linked credit on short term. Under this arrangement, the creditors are themselves traders and large landholders who borrow from commercial banks. To avoid charging interest payment, local traders purchase crops at price lower than the market level, and they require pledging animals as collateral for advancing credit. Ali (1986, p. 343) estimated that this informal system of transaction amounted to a payment of disguized interest rates, ranging from 60 to 280 percent! The *sheil* system of borrowing has, therefore, led to a considerable transfer of income from small farmers to traders, and to many small farmers sinking into chronic indebtedness.

The Impact of Governing the Rural Economy on Food Production

The discussion turns to examine the food production consequences of state policies outlined in the preceding sections. Given the important weight of agriculture in the national economy indicated in Chapter 3, the increasing role of the central government in agricultural production and trade has greatly influenced total GDP, food security and the level of living of rural population.

Falling Productivity in a Stagnating Rural Economy

The macro-economic indicators suggest that agricultural GDP has stagnated during the last three decades. According to estimates made by Sudan's Ministry of Finance and Economic Planning, the volume of agricultural GDP at 1981/1982 fixed prices remained virtually unchanged between 1973 and 1987 at an average £S 2,100 million. At the end of the period, it was 97 percent of its average level in 1973-1975. With a policy discriminating against agricultural development in Southern Sudan,[24] and in the vast traditional sector of Northern Sudan, agricultural output (crops and animals) also stagnated during the longer period 1960-1989, fluctuating around the low annual growth rate of 1.5 percent.

This is not surprising in a deteriorating economy whose public and private investment was falling. Public investment in agriculture fell from an average of 35 percent to 25 percent during this period, as a part of the agreement with IMF and the World Bank implemented in 1980-1986 and which required substantial cuts in government expenditure. The creation of the private free foreign-exchange market in 1981 led to the flight of private savings out of the country, estimated at US$ 14 billion between 1981 and 1986.[25] These public and private actions have resulted in a low share of total investment in GDP of 8 percent on average, compared to the World Bank's weighted average for the world's low income-economies of 26 percent during the same period.

There has been a downward trend in both productivity per working person in agriculture and in food production per head of total population. Total agricultural and food production failed to keep pace with growth rates of agricultural labor force and those of total population, respectively. Relative to the base period 1979-1981 = 100, the annual rate of growth of labor productivity declined steadily from 1.9 percent in the 1960s to 0.5 percent in the 1970s, and fell further to 0.1 in the 1980s. Alarmingly, the annual rates of change in food productivity fell to a negative 1.6 percent during the period 1980-1989 (1990 *FAO Country Tables*).

Cereal Production Instability. Not only did total agricultural output stagnate, but also there was a high degree of year-to-year cereal production instability, with falling yields of sorghum and millet between 1948 and 1990 (see Table 6.5 and Appendix Table 1). Such a disturbing situation adversely affects not only human nutrition through cereal availability for consumption, but also agricultural growth, productivity and rural employment (demand for labor and wages). Likewise, it increases Sudan's dependence on food aid. The main factor behind this extreme inter-annual instability is the reliance on the highly variable rainfall. This reliance has resulted from the very low rate of irrigation expansion during the last three decades, from 14 to 15 percent between 1960 and 1990.

The noted earlier fall in budgetary allocation for agriculture (including irrigation) at the sectoral level is, however, misleading. Disaggregated data by rural localities show, for example, that the two Western provinces of Kordofan and Darfur, which produce most of the sorghum and millet under rain-fed agriculture, received only 9 percent of total allocation to agriculture in the Six-Year Development Plan, 1977-1983, of which 60 percent was actually spent. These two provinces have frequently suffered from prolonged droughts, leading to famine. They account for almost half the total area of pastureland, supporting the nomads and their animals. The two provinces also comprize 30 percent of total rural population, nearly 53 percent of nomadic pastoralists and 42 percent of economically active population in rural Sudan (Farah Adam, 1981 and *Statistical Abstracts*).

TABLE 6.5 Fluctuation in Cereal Areas and Yields in Sudan, 1948-1990

| | Area Thousand Hectares[a] | | | | Yield Tonnes per Hectare | | | |
Years	Sorghum	Millet	Maize	Wheat	Sorghum	Millet	Maize	Wheat
1948-1952	820	332	14	13	0.7	0.5	0.9	1.2
1961-1965	1,400	520	26	27	0.9	0.6	0.7	1.3
1969-1971	1,828	750	39	118	0.8	0.7	0.8	1.1
1979-1981	3,163	1,094	67	206	0.7	0.4	0.6	1.0
1983-1987	4,390	1,420	52	120	0.6	0.6	0.5	1.4
1988	5,813	2,320	40	144	0.7	0.2	0.7	1.2
1989	3,801	1,569	35	199	0.4	0.1	0.7	1.2
1990	2,905	1,150	50	258	0.5	0.1	0.7	1.5

[a] average of harvested area for each sub-period.

Source: *Production Yearbook*, Food and Agriculture Organization of the United Nations (FAO, Rome), several issues.

The sharp fluctuation in cereal production in rain-fed areas, from one year to the next is the main source of falling labor productivity and food insecurity, at both national and household levels (see Figure in Appendix 1). Based on Hussain's index for national food security (sorghum, millet and wheat) during the period 1978-1987 and which excludes cereal foreign grants, the degree of insecurity (failure to meet total consumption requirements) ranged between 14 percent in 1982/1983 and 53 percent in 1984/1985 (Hussain, 1991, Table 5.1). By including foreign aid, Maxwell (1989) found that Sudan was self-sufficient from all sources of cereal supply between 1971 and 1984, but not in 1985-1987. Thus, Sudan has been unable to meet the minimum food needs of its people from its own resources.

On the demand side, the influx of refugees and people displaced by the war in the South (roughly about 2 million persons in 1988) has aggravated the food situation. Moreover, rapid urbanization at the average annual rate of 5.4 percent in 1960-1990, and changing diet preference of urban population and rich farmers (substituting wheat for sorghum) have increased the aggregate demand for wheat (bread) that added a new dimension to food security problems. Despite the dramatic expansion in the area growing wheat, particularly in irrigated Gezira at the expense of cotton area, imports of wheat both grain and flour have increased fourfold between 1960 and 1990 (FAO *Country Tables*, 1990). After the government authorized growing wheat in irrigated land, including the Gezira scheme, harvested area of wheat increased from an average of 71,000 ha. in 1965-1968 to 147,300 ha. in 1971-1975, and it rose further to 161,000 ha. in 1981-1985. Consequently, cotton area fell in the latter two periods from an average 502,000 ha. to 390,000 ha. In the allocation of their irrigated land between the two crops, farmers had no incentive to grow cotton, whose price was depressed by government monopoly and costs of its production was rising. The cultivation of wheat, on the other hand was attractive: it provided food security to the farmers' households; the demand for wheat was rapidly rising; and its farmgate price was supported, and bread subsidized.

As one would expect, substituting wheat for cotton in irrigated land has given rise to controversy over comparative advantage in resource use, on the one hand, and socio-political questions of food security, reliance on wheat grants and Sudan's foreign relations with donor countries, on the other. Apart from competing with wheat in the use of scarce irrigation water during the dry winter season, cotton is more profitable to the economy; its gross value of output per feddan, at competitive world prices was approximately ten times as much as that yielded by wheat.[26]

Widening Inequality and Persisting Poverty

The pattern of capital investment favoring the modern sector has influenced the use and remuneration of rural labor, and has generated inequality in the size distribution of landholdings and income between and within modern and traditional sectors.[27] Further adverse income distributional effects have resulted from the introduction of price and foreign-exchange liberalization policy, in the early 1980s, as a part of the World Bank and IMF-induced structural adjustment program. Whereas it removed barriers to stimulate private investment and trade, and it reduced distortions in crop prices, the program has greatly favored the already rich traders and the producers of commodities for export.

Inequality in the distribution of income and the erosion of the poor's purchasing power have, since the 1970s, increased under inflationary pressure rising annually at the average rate of 33 percent in the 1980s. According to the findings of various studies (ILO, 1986, IDS, 1988, Diab, 1989 and Hussain, 1991) cost of living has increased, and real wages as well as purchasing power of the poor eroded. Rising food prices hit the net-buyers of food among poor farmers and landless laborers harder than the rich. Moreover, continuing mechanization of agriculture tends to reduce the demand for labor.

To understand the distributional changes in rural Sudan, we need to examine the size distribution of landholdings, and the results of nutrition and rural households' incomes or expenditure surveys. This broad frame helps to understand a good part of the process of poverty-making. There are, however, wide gaps in official statistics. What exist are incomplete surveys of landholdings and household budget surveys that *exclude* the Southern provinces, the nomadic people and the size distribution of animal wealth.

Skewed Distribution of Landholdings

A study carried out in 1980-1981 by the Ministry of Finance and Planning in three provinces (Gezira, North Kordofan and Southern Darfur) reveals that 91 percent of the surveyed holdings (*hawāsha*) were below 4 hectares each. The distribution in rainfed North Kordofan was very unequal; 51 percent of holdings were less than 4 ha. each and 21 percent over 15 ha. each (Abu-Sheikha, 1983, Table 29). In the same year, a detailed sample survey of 296 rural households in al-Gedaref rainfed district of Kassala Province was carried out in 1984 by the International Labor Organization of the United Nations (ILO). The findings show a skewed distribution of landholdings: 23 percent of surveyed households held land in the size group of 4 hectares, 73 percent had, on average, 100 hectares each, and the rest or 4 percent were in the top size group of over 200 hectares (ILO, 1984, Tables 2.6 and 2.7).

These scattered data which invariably left out Southern Sudan, suggest that a very considerable variation in the degree of inequality exists among and within the provinces. Part of this inequality has been generated by a deliberate government policy on land allocation among the lease-holders in the modern sector. As explained earlier in the section on modern irrigated sector, the government allotted each tenant a unit of 17 hectares in the Gezira scheme, reduced to 6 hectares in Managil and Khashm el-Girba, and it was lowered further to 4 ha. in government-administered pump schemes. Likewise, in the mechanized rain-fed scheme, the average unit rented out by the state-run Mechanized Farming Corporation was 625 hectares and, in some cases, the influential enterpreneurs were able to rent-in 5 and even 20 demarcated units, ranging from 3,000 to 14,000 hectares each (ILO, 1984, p. 27). In some localities, district and local government authorities turned a blind eye to entrepreneurs who have encroached on the pastoralists' lands, pushing them and their livestock to less suitable grazing areas.

Incidence of Rural Poverty

The lack of reliable data on income distribution, disaggregated into rural and urban, is a serious constraint in understanding the dimensions of poverty and estimating its incidence. The only sources of relevant data are the household expenditure survey (HES) which was conducted in 1978/1979 and the nutrition survey (Sudan Emergency Recovery Information Surveillance System, SERIS), conducted in 1986/1987. Both surveys cover the 14 provinces of Northern Sudan, and exclude the seven provinces of the South.

In 1983, and based on the preliminary results of the 1978/1979 HES, an attempt was made to quantify the poverty conditions in rural areas. A team of Sudanese economists, nutritionalists and a few professionals from government departments concerned with the subject joined Abu-Sheikha, an expert from FAO to estimate the extent of rural poverty. They established an income-based poverty line at £S 500 for a rural family of six persons. Accordingly, it was estimated that about 70 percent of rural population were living in absolute poverty. Though it is an arbitrary estimate, its extent is appalling.

The team of researchers identified the determinants of rural poverty as a combination of: (a) lack of adequate access to land, water, credit and marketing assistance; (b) limited ownership of non-land assets; (c) exploitative production relations between private irrigation pump-owners and small tenants; (f) indebtedness of small farmers caused by the system of *sheil* for crop marketing-linked credit; and (g) low human physical energy owing to undernutrition and ill health which, in turn, have lowered their productive ability (Abu-Sheikha, 1983, pp. 91-95). These poverty conditions were reflected in the high illiteracy rate of 80-85 percent, and a high infant mortality being 58 percent above the weighted average of the world's low income economies in 1983. Moreover, low life expectancy at birth was very low at 47 years for female and 49 years for male; 23 percent lower than the weighted average of all the 35 low income economies classified by the World Bank's development indicators for the same year.

Through a social accounting matrix, Diab and Evans (1991) used the results of the 1978/1979 HES to link food security to income expenditure levels by size income groups and by rural/urban classification. Other economic variables that reflect income distribution and agricultural growth policies were also linked in this analysis. They found that all rural households in the category of less than £S 500 and most of those in the income size group of £S 500 - 999 were food insecure. Accordingly, about 58 percent of all rural households were estimated to be unable to have the minimum income to secure the needed food for survival. If we allow for an additional minimum level of non-food goods and services, the incidence of rural poverty may roughly reach 75 percent. This is still lower than the 85 percent estimated by the World Bank.

As nutritional status is a fundamental characteristic of poverty, we consider also the anthropometric measurements of undernutrition among a sample of 80,000 children under 5 surveyed in 1986/1987 (SERIS). Although it was a good agricultural year, the survey results show that the severely undernourished (less than 80% of standard weight for height) represented 13 percent of children surveyed in rural (sedentary) areas, and 14 percent among the children of nomadic people.

Vulnerability of the Poor to Famine

Regardless of the precise figure of the incidence of poverty, the approximate magnitude of its various estimates and undernutrition levels is so staggering that it stirs one's mind.

The vulnerability of the poor to the disastrous consequences of prolonged droughts is greatest, compared to the non-poor farmers. The effects of the low rainfall in 1981-1982 followed by the severe drought of 1984-1985 which brought about the 'famine that kills' (to use the idiom of de Wall, 1989) have been documented. Whereas a failure of agricultural products was caused by a severe drought, famine is not just a result of drought. Nor does drought come as a surprise. Based on his analysis of rainfall records over a period of 30 years (1951 - 1980), a Sudanese authority on the subject asserts that the extrapolation of the trend in rainfall was found to be consistent -- at 80-90 percent probability --with the very recent increase in the severity of drought (Abbās Abdalla, 1986, p. 309 and Table 16.2).

The sad findings of the studies on both famines of 1939 in the Equatoria province, and of 1984-1985 in Darfur and Kordofan provinces are striking. They reveal how the real causes of poverty in Sudan, which remained dormant and of little concern to public media, have been brought to the surface by the recurrence of famine in the 1980s. They suggest that the institutional setting in rural areas, including the inadequate access to health services and the bureaucratic establishment in local government is an important contributing factor. Temporary relief work to deal with the outcome -- and not an assault on the determinants of poverty and causes of mortality in famine -- has been the government authorities' dominant approach to these famines.

The 1939 famine was viewed by the, then, British administration as a natural punishment of small-scale farmers in the Equatoria province for not obeying the orders issued for "their real interests" to grow cotton instead of sorghum in order that they had sufficient cash to pay the poll tax, and to buy grain when needed. The, then, Governor of Equatoria in Juba considered it unwise to provide relief aid because these poor farmers "had refused or failed to take necessary steps to protect themselves against famine".[28] In the 1984-1985 famine, thanks to the generous relief work of the USAID, EEC, United Nations' agencies and foreign non-governmental organizations, the approach was more sympathetic than that in 1939 (see details of the relief program in de Waal, 1989, Chapter 8).

Whereas the foreign agencies saw the issues of famine in terms of a short-term shortage of food and transitory undernutrition, the Sudanese bureaucratic establishment considered it as a passing emergency, and concentrated on the organization of charity for the survivals. In tackling famine, the official action was a coordination of Islamic almsgiving *'Zakāt'* and the distribution of foreign aid for the sustenance of the survivals. But all these forms of aid were temporary, and concentrated in towns. De Waal (1989 and 1991) and Keen (1991) reported that negligible aid had reached the needy nomadic communities and nearly one-third of food distributed was leaked to local grain traders. The relief aid and accompanying enthusiasm stopped after the famine. Those who survived relied on their diminished resource-base for survival.

In Darfur alone, the human deaths were excessive; nearly three times a normal year. The excess was estimated at 95,000 persons who died together with about 55 percent of their livestock.[29] Furthermore, 35.6 percent of the children in Kordofan province suffered severe undernutrition (Oxfam and UNICEF estimate, June 1985).

The effects of the 1984/1985 famine lasted for several years. In 1991, it was found in North Darfur and North Kordofan that those households who lost their animals had not replaced them, and that the children's undernutrition (less than 80 percent of weight for height) has become chronic. The nutrition survey conducted by the Ministry of Health in these two provinces from December 1990 to January 1991 reveals a high incidence of 20.5 percent in North Darfur and 18 percent in North Kordofan (FAO/WFP Mission report, 1991).

All these brutal sufferings and causal antecedents of high mortality in famine seem to have not brought about fundamental changes in: the rigid land tenure system, the pattern of agricultural growth, and the design of rural development strategy. The evidence from the various studies is unequivocal: the rural poor and victims of famine were the food-insecure small-scale farmers, poor nomads, landless hired workers, widows and female-headed households in the traditional rain-fed sector. One would have expected in the recently implemented structural adjustment policy an anti-poverty and anti-food insecurity rural development policy targeted to these groups. One would also have expected a change in the path to agricultural development to translate the growth benefits into productive gains for the economically viable poor, and to take public action for famine prevention. These necessary policy changes are needed to enable the millions of the rural poor to escape the risk of future prolonged droughts and famine. A firm political commitment is needed to realize the potentials of rural Sudan and the aspirations of its poor.

Notes

1. *Uhda* means a number of villages for which each high official or influential family had the responsibility to collect taxes. *Shiflic* and *Ibadeya* denote tax free, large areas of land granted to members of the ruling family and to notables. For a detailed description of these types of arrangements, see Baer, 1962, pp. 11-19.

2. This statement refers to Land Taxation scheme approved by the Cabinet of Ministers on 30 July 1946 and its accompanying note prepared by Land Taxation Department (in Arabic).

3. Kentar is a measure of weight (*Zahr*) equals 157 Kilograms including seeds and 50Kg without seeds (*Sha'r*).

4. Despite these socially undesirable actions, the British Administration were behind the issue of the Five Feddans Law of 1912 sponsored by the Viceroy Lord Kitchener following the widespread loss of land by small farmers resulting from the economic crisis of 1907-1908. Although the law was not effectively implemented, it exempted

land-owners of 2 hectares and less from foreclosure against debt. This policy was followed by the sale of a relatively small area of 15,000 ha. of state land in Fayoum province to small farmers. Another positive measure was the establishment of the Aswan Dam in 1902, which contributed to an increase in cultivated area by 73,000 ha. However, this expansion was offset to a large degree by the granting of 30,000 ha. by the Egyptian Government to become a British military base in Suez and Ismailia provinces.

5. Ministry of Finance, Public Domains Department, *"Maslahat Al-Amlak Al-Amereya Inshā'oha wa A'amāloha"*, 1949. This report shows that between 1939 and 1949, out of total state-owned land sold, large landowners purchased 90.7%, agricultural colleges graduates 7.6% and only 1.7% of the *fellaheen* were land recipients.

6. By several laws issued between 1956 and 1960 cotton area was limited to one third of each holding and wheat to another third. This allocation would consequently regulate growing other crops in the crop rotation.

7. The shift in policy started in 1973 by changing the priority for the eligible categories to receive reclaimed lands whereby government employees and members of the armed forces were given priority over the landless agricultural workers and small tenants. This change was provided in the ministerial decree no. 105 of 1970 and number 59 of 1975. See also recent shifts in the report of the Committee on Agriculture of the Parliament, the summary of which is published in al-Ahram, 21 February 1990, p. 3, in Arabic. The political parties' debate on returning the determination of land rent to market forces is reported in (*Al-Ahram*, 13 January, 6 and 14 February 1992, and *al-Wafd*, 24 January and 21 February 1992). See also my own analysis of the underlying issues on economic grounds, *"Qalaq Al-Fallah* (the *Fellaheen's* anxiety and uncertainty)", *al-Ahram Al-Iqtissadi*, 23 September 1991, in Arabic.

8. These changes are based on data given in the Ministry of Agriculture "Bulletin of Agricultural Economics" (*Nashrat Al-Iqtisad Al-Zira'ee*) 1960 and 1968, and *Statistical Yearbook*, CAPMAS, June 1988.

9. This study was based on a 5% random sample survey of the beneficiaries' households in three land reform cooperative districts (Inshas, Itai Al-barood and Gabaris). The results are presented in Chapter 6 of El-Ghonemy, 1990.

10. See results of micro-studies in: Gadalla (1962): Saab, (1967); The Study prepared by Land Reform Authority, Ministry of Agriculture, June 1966; and Abdel-Khalek, Mohsen, (1969).

11. There are different estimates of returning migrant workers. See, Simon Commander (1987) and El-Firgany (1988).

12. The estimation method employed is Ordinary Least Square multiple regression with rental values as the dependent variable (R). Cultivable land *per capita* rural population i.e. the inverse of conventional density (D), cotton price (Pc), value of gross output per feddan of cropped land (Y), and a dummy variable for state authority over rental values (G) are independent variables. Therefore the estimated relationship is

$$R_t = a_0 + a_1 Pc_t + a_2 Y_t + a_3 G + e_t$$

where e_t is the error term. The relationship is estimated separately for three periods: the whole period (1913-1986), the period up to 1952, and for the period of rent control 1955-1986. It was found that the density variable which is not shown in Figure 6.1, but given in Table 6.2 has the expected sign during 1913-1952, but changed to opposite sign after government intervention (1952-1986), which, on economic grounds, is contrary to expectation. Thus, two important variables (density and the value productivity of the scarce factor in agricultural production, land, appear to have lost their economic importance in the face of equity-orientated state control. See the analysis and results of estimation in El-Ghonemy, 1992, pp. 181-4.

13. The reliance of the rural poor on food subsidies for their daily consumption is high. In 1984, Harrold Alderman and Joachim von Braun of the International Food Policy Research Institute (IFPRI) identified the beneficiaries of the extensive subsidy system. They found that: (a) 17.9% of food expenditure of the poorest quartile among rural households came from subsidized food items; (b) 50% of total calorie-intake came from government supply shops (*Tamween*) and co-operatives, representing 16% of total and 34% from subsidized bread and wheat flour purchased; (c) this botton quartile of rural households consumed 2,337 total calorie intake *per capita* per day compared to average 2,654 calories in all rural households and 3,163 calories *per capita* of total population; and (d) subsidized cereals consumed by small landholders of less than one feddan each represent 40% of total cereals consumed per person in one year. Subsidized cereals were mostly bread, rice and wheat flour.

14. See for example the arguments made by Farah Adam and Hashim Awad in Shaaeldin, editor, (1987), and El-Hassan and Ahmad Ali in Barnett *et al*, editor (1988).

15. Abd Ar-Rahim (1990: p. 756) and Hashim Awad (1987: pp. 33-4).

16. Officially called the Anglo-Egyptian Condominium, the Sudanese Historian Abd ar-Rahim says "it was nominally Anglo-Egyptian but was actually a British colonial Administration" (1990: p. 757). Another Sudanese University Professor states that "Condominium rule was essentially a British rule; Egypt itself being a protectorate of the British Crown" Ahmed Ali (1988: p. 24).

17. The account of the situation leading to the establishment of the Gezira Scheme is based on Gaitskell (1959).

18. See Mahgoub "Land Policy and Settlement in Sudan" in El-Ghonemy, editor (1967, pp. 175-188) and Shaw study in Shaaeldin, editor, (1987).

19. See the results of the study conducted by Shaw, p. 147 and Table IV in Shaaeldin. Working women in agriculture is from National Development Plan 1978-1983.

20. Ministry of Agriculture, *Statistical Bulletin*, several issues.

21. For a detailed account of planned land settlement schemes, see Mahgoub, *op cit*.

22. In 1981, the leader of the National Islamic Front, Dr Hassan Turābi was appointed Attorney General in the Cabinet of President Nimeiri and his deputy, Abd al-Rhaman Mohamad, Minister of Interior. Following several changes in government, before and after Nimeiri's removal from office, Dr. Turābi became Deputy Prime Minister in February 1989. He and his Front insisted all the time that Islamic Law should be implemented, without any compromise with the demands of the South. This hard line position had prevented reaching an agreement for national unity between the central government and the leaders of the South.

23. See an explanation of these procedures in Chapter 2 and its note 3.

24. For a detailed account of the discrimination against the development of agriculture in the South, see Albino, O. *The Sudan: A Southern Point of View*, Oxford, 1970, pp. 90 - 93. On the neglect of the traditional sector, in the 1960s, see Osman, A. 'Planning Agricultural Development and Policy Implementation', unpublished Ph.D. dissertation,

Michigan State University, 1969. He states that of the total planned investment in agriculture (1957-1969), the share of traditional agriculture was only 7% (p. 69).

25. The fall in the share of agriculture in total public expenditure is from the Ministry of Finance and Economic Planning. The capital flight is from Hussain, 1991, p. 106. The share of investment in GDP is calculated from the World Bank development indicators.

26. On this controversy, see *IMF Staff Papers*, "The Experience of Sudan", March 1980, Hussein (1991), and the World Bank, *Sudan Export Development*, No. 4263, 1983.

27. For understanding the resulting differentiation in the demand for labor between irrigated and rain-fed sectors, see ILO study, *Labor Markets in the Sudan*, 1984.

28. These quotations are taken from a letter to the district commissioner in Torit, dated 2 November 1939, annexed to Hodnebo's study, 1981.

29. De Waal made this estimate, which is based on the difference between crude death rate in normal year and the famine period (June-December 1984 till December 1985). See de Waal 1989, Table 7.1 for human deaths and pp.152-3 for livestock deaths. In their study on the effects of drought and famine (1983-1988) on livestock in Western Sudan, a team of researchers from IFPRI, estimated that 71 percent of the cattle and 41 percent of the goats were lost in North Darfur (IFPRI 1991, p. 35)

7

Growth and Equity in Rural Development: An Inter-country Comparison

For understanding inter-country changes, the discussion is based on two beliefs. One is that the North African nations prefer sustainable rural development to stagnated rural economy. The other is that they prefer a rapid poverty reduction and greater redistributive equity to rising poverty and inequalities of opportunities. Both beliefs are linked by way of the economic organization of agriculture, rural institutions and the political process. They are preferred social choices, differently expressed in the countries' development plans.

The first section of this chapter discusses the sources of inter-country variations in productivitgy and growth rates of total agricultural output and food production, with special emphasis on technological change. In particular, it discusses why in a few countries sustained agricultural growth has taken place, why and where it has not, and the different degrees of distributive equity which the patterns of growth have produced. The second section examines the effects which the fragmentation of farm size have on food production, and addresses the question: are Islamic principles of inheritance to blame? The subsequent sections examine the variations in land concentration, poverty levels and quality of life, relating them to agricultural growth.

Technological Change and Agricultural Growth

Consider Table 7.1 which presents the variations among the six countries in growth levels and in three principal technological changes (irrigation, chemical fertilizers and farm machinery). There are factors other than technology which influence the variation in output growth. They include climatic fluctuations, land tenure arrangements, access to credit and marketing facilities, and the extent of government intervention in market mechanism; the production impact of which was discussed in the preceding chapters. This section, therefore, concentrates on the countries' varied investment rates in technological change and the resulting contribution to agricultural growth.

As established conceptually by Harrod-Domar in the 1940s, there is an important association between technological change through capital investment and output growth, leading to economic stability. By capital investment in agriculture is meant gross investment, both private and public. It is estimated for individual countries by the United Nations as a part of its calculation of national accounts, and expressed in terms of comparable US dollars, known as gross fixed capital formation (GFCF). The principal components of capital investment in agriculture are irrigation, land reclamation, drainage, soil conservation, farm mechanization, food storage, and public current expenditure on inputs' subsidy, agricultural research and extension services. However, current expenditure is excluded from GFCF estimation.

In principle, technological change induces commercialization of agriculture, stimulates output growth, and reduces uncertainty and the degree of food production instability. Given the North African climatic conditions, irrigation is a major contributor to increased agricultural output, and sharply reduces the risk of crop failure. These favorable effects of technological change are reflected in crop yields per unit of land and agricultural GDP per working person in agriculture. Depending on technology choice, technological change also improves employment opportunities in rural areas, and affects the distribution of growth gains by farm size. These positive effects are central to rural development, but they may or may not benefit poor producers and landless workers.

Irrigation Expansion and Land Development

In North Africa, irrigation absorbs approximately 40-50 percent of public-sector investment in agriculture. The quantum invested depends, therefore, upon the share of agriculture in total investment, including public-capital expenditure.

Before making inter-country comparisons, three principles need to be stated. Firstly, the multiple effects of capital expenditure for irrigation expansion on production growth do not all occur at once. There is a time lag from 5 to 8 years between actual capital expenditure and realized output growth. Different characteristics of soil require different periods for bringing productivity up to the desired level. Secondly, at the aggregate level, the net quantity of actually cropped area may be less than the sum of cultivable old area and the newly reclaimed area because of rapid urbanization and the rates of taking land out of production for non-agricultural purposes. Thirdly, the data on gross investment

TABLE 7.1   Investment in Agriculture per Units of Land and Labor Force, and the Rates of Growth in the Six Countries, 1960 - 1990

Category	period	Algeria	Egypt	Libya	Morocco	Sudan	Tunisia
1. Gross Investment in Agric.[a]							
As a share of total public exp.	1980-1985	11	9	13	7	10	13
The ratio of the share to agric. GDP as % of total GDP[b]		1.8	0.4	4.3	0.4	0.3	0.8
GFCF per agric. laborer US$[c]	1978	n.a	45	4,464	n.a	n.a	240
	1988	n.a	238	3,933	n.a	n.a	473
" " per arable ha. US$[c]	1978	n.a	86	244	n.a	n.a	30
	1988	n.a	495	274	n.a	8	67
2. Irrigation Expansion							
Irrig. area (thousand ha.)	1960	229	2,568	121	875	1,480	65
	1985	338	2,486	234	1245	1,848	241
Percentage change	1960-1985	+77	-3	+93	+42	+25	+270
3. Fertilizers Used							
Kilograms per arable ha.	1960	7	94	2	4	2	4
	1985	38	347	28	36	7	19
Percentage increase	1960-1985	442	269	1,300	800	250	375
4. Farm Machinery							
No. tractors per 1,000 ha.	1960	4	5	1	1	n.a	2
	1985	10	17	13	4	1	5
Percentage increase	1960-1985	150	240	1,200	300	-	150
Harvesters & Threshers							
No. in use	1974-1976	3,850	1,927	n.a	2,708	900	2,343
	1987	8,628	2,243	n.a	4,570	1,200	2,600
Percentage increase	1974-1987	124	16		69	33	11
5. Agric. Growth Annual Rates %							
Total agric. production[d]	1960-1970	1.8	3.0	7.3	4.5	2.5	0.1
	1980-1989	3.6	2.8	4.1	6.8	-0.1	2.5
Food production[d]	1960-1970	1.8	3.1	7.3	4.5	2.1	n.a
	1980-1989	3.5	3.6	4.1	6.8	-0.1	2.5
Agric. GDP	1980-1990	4.3	2.5	n.a	6.4	n.a	2.3

Notes: n.a stands for not available. a. See definition in the text. b. This is the ratio of percentage allocation of public expenditure to agriculture divided by the percentage share of agricultural GDP in total GDP. c. See definition of GFCF in the text. d. Based upon 1979-1981 = 100.

(continues)

Table 7.1 (continued)

Sources: The share of investment in agric. to total public expenditure is taken from IMF *Government Finance Statistics, Yearbook*, various years, countries' development plans and *Atlas of African Agriculture*, FAO, Rome, 1986. 1.b is calculated by the author from the share of investment in agriculture to total. The share of agric. GDP in total GDP is from *World Development Report*, several years, Oxford University Press. 1.c is from *The State of Food and Agriculture*, 1981, 1989 and 1991, Annex Tables: FAO, Rome. Rows 2,3,4 and 5.d are from *Production Yearbook*, several years and 1991 *Country Tables: Basic Data on the Agricultural Sector*, FAO, Rome. Agricultural GDP average annual growth is from the World Bank, *World Development Report*, 1992, Indicators, Table 2.

in agriculture are underestimated; they do not include non-monetized investment of the farmers' own labor and the efforts of nomadic people in raising livestock.

We need also to bear in mind that definitions of what constitutes public expenditure on agriculture vary from country to country, and therefore, their shares are not perfectly comparable. Moreover, they preclude private investment, about which reliable data are scarce. Scattered information suggests an approximate share of 20-30 percent of gross investment in agriculture. However, and considering the governments extensive control of resource use in agriculture between 1960 and mid-1980s, public-capital expenditure was dominant. This dominant role was exercised in irrigation, drainage, land reclamation schemes and also in the importation and supply of technical inputs.

Given these elementary principles, and related caveats, Table 7.1 provides a perspective on inter-country variations in gross investment in agriculture. In particular, it shows the wide range of changes in public expenditure, and the resulting changes in irrigated areas. The ratios (row 1.b), together with the data on GFCF per unit of land and agricultural labor give an approximate indication of the countries' degree of commitment to agricultural development. Libya and Algeria, the two countries with the highest GNP per head and oil-export revenues allocated a higher share in public expenditure to agriculture than its low share in total GDP, the ratio being 4.3 and 1.8, respectively. At the other end of the scale, Egypt, Morocco and Sudan with their much lower GNP per head allocated disproportionately low shares to agriculture relative to its contribution to total GDP, the ratio being 0.4, 0.4 and 0.3, respectively. Tunisia, a middle-income country, has virtually maintained a balance between the two shares (the ratio was 0.8 in 1980-1985).

Table 7.1 also shows that from 1960 to 1985, the most marked *proportionate* increase in irrigated area, was in Tunisia followed by Libya, Algeria and Morocco. The low rate of expansion in Sudan is not unexpected, considering its lowest level of GNP per person and the ratio. We have explained in Chapter 6 the causes of Egypt's negative net balance of the effective supply of land (see Table 6.1 and p. 100, and El-Ghonemy, 1992). We may also recall from the discussion in Chapter 3 (Table 3.4), how total food and cereal production instability is inversely related to the proportion of the area of irrigated land to total arable land. During the period 1969-1984, the instability index was the lowest in Egypt (at 1.9) where irrigated area was 97 percent of total, compared to an index of 19.7 for Algeria whose percentage of irrigated land was only 5.

The Impact of Government Expenditure Instability on Agricultural Growth

The preceding discussion and the data given in Table 7.1 suggest that the countries which achieved the highest growth in irrigated area during the period 1960-1985 were also able to realize the fastest agricultural growth (Libya, Tunisia, Morocco and Algeria).

As noted in Chapter 6, Egypt had already realized a complete irigation fo field crops. Its slow growth lies, not in climatic fluctuations, but in policy shifts. To illustrate the relationship between the instability in government expenditure and output growth, the data of Egypt and Libya are presented. In Egypt, agriculture (including irrigation) received 23.4 percent of total public-capital expenditure in 1959-1964, during which the average annual expansion in irrigated land was 48,330 hectares. When agricultural share was reduced in 1965-1975 to 14.8 percent, expanded area fell to 10,000 ha. per year. Subsequently, that share fell further to 11 percent in 1976-1985 and the average annual increase in reclaimed, irrigated land also fell sharply to 6,958 ha. During these three periods, agricultural GDP annual growth rate fell also from 3.3 percent to 1.3 and 1.9 percent respectively (Ministry of Planning and *Statistical Yearbook*, 1981 and 1986).

The Libyan situation resembles the Egyptian experience with regard to the instability of the public-sector's expenditure (on agriculture) on output growth *via* fluctuated investment in expanding irrigated area. While fully irrigated land almost doubled between 1960 and 1985, expanded area together with agricultural GDP have closely followed the movement in capital expenditure. The latter was subject to the fluctuated flow of oil-export revenue. When agriculture received one-tenth of total development expenditure in 1964-1970, the area of irrigated land expanded at the annual rate of 6,700 ha. and both total agricultural and food production grew at 7.3 percent *per annum* (FAO *Country Tables*, 1991). As a result of the sharp rise in oil revenues in the 1970s, the rates of the other three variables accelerated, and agricultural GDP annual growth rate reached a high record of 11.8 percent compared to an average 2.7 percent in the Middle East and 2.9 percent in Sub-Saharan Africa. In the 1980s, government oil-revenues fell; so did the proportion of capital allocated to agriculture (from 17% to 13%). At the same time, newly irrigated land declined from 7,000 to 2,800 ha. per year. Consequently, agricultural growth fell from 11.8 percent to an annual rate of 7.3 percent in 1980-1985 (see Tables 3.1 and 5.4). As shown in Table 7.1, food production growth also slowed down in Libya in 1980-1989.

Technological Change in Food Production

The countries' investment efforts to expand irrigated areas has, since the 1960s, induced the introduction of technological change in food production in response to demand changes. Although irrigation is necessary to minimize instability of food production, it is not sufficient to increase output and farmers' income. Other important technologies include: the adoption of high-yielding varieties of cereals and vegetables seeds, intensified use of fertilizers, appropriate type of mechanization, drainage works, improved livestock production and health, and effective agricultural research and extension services.

The incomplete set of available data makes it difficult to quantify the impact of each of these technologies on food production which is the prime interest of this study. Instead, we look at the effects of the adopted technological package on inter-country variations in yields of cereal crops and major vegetables, and kilograms of meat and milk per head of livestock given in Table 7.2.

Variation in the Response of Yields to Technology. The six countries have, since 1970, intensified the use of chemical fertilizers and high-yielding seeds. Both land area-saving and yield-increasing inputs were heavily subsidized by governments until their gradual removal in the late 1980's after the introduction of fiscal reforms and agricultural trade liberalization. Proportionally, Libya and Tunisia have increased yields of all cereals, in general, and bread wheat, in particular, at rates higher than the other four countries during the period 1961-1985. This variation reflects both the level of GFCF per working person in agriculture and the scale of irrigation-expansion shown in Table 7.1. Seemingly, in Algeria, Sudan and Morocco, the disproportionate low investment in the large traditional rainfed-sector that produces most of the cereals, may partly explain their slow increase in yields. Price policy has also its effects. In Morocco, a field study in Abda Region of Safi Province, found that most fertilizer was used in irrigated agriculture, and that the government-administered low wheat price did not warrant the adoption of improved seeds during the agricultural year 1985-1986 (Abbott *et al*, 1991, pp. 28-30).

Egypt's problem is different. In spite of its unique advantages of soil fertility and perennial irrigation in 97 per cent of cereal land, the high average quantum of fertilizer use per hectare has, since the 1960s, brought about a modest increase in yields. As explained earlier (Chapter 6), distorted pricing of wheat together with falling agricultural investment by the dominant public-sector are contributing factors behind the low rise in the already high average level of yields.[1] Moreover, the Egyptian government's sluggish investment in drainage works has led to the deterioration of the over-irrigated land, resulting in the degradation of 22.3 percent of total cultivated area in 1980, from first and second classes of soil quality down to third and fourth classes. Furthermore, farm areas using high-yielding seeds of wheat, maize and rice fell from 71 percent of total in 1970 to 53 percent in 1980.[2]

With regard to Algeria, yields of food crops, except sugar beetroot, did not respond to the relatively high investment in agriculture, particularly the 4-fold increase in fertilizer use between 1961 and 1985. This may be due to the stagnated subsector of production co-operatives (collectives) which relied heavily on government transfers in order to reduce losses, arising from low output and lack of incentives among the *fellaheen*. Until the introduction of institutional and economic reforms in 1985-1987, these government-controlled co-operatives were dominant in rural Algeria and inefficiently managed (see Chapter 4, pp. 52-54 and Appendix 3).

Vegetables' Contribution to Agricultural Growth. All published statistics on agriculture in North Africa point to the remarkable increase in both areas and yields of vegetables, the output of which has been growing faster than total agricultural output. Vegetables in Egypt and almost 75 per cent in Tunisia are cultivated in irrigated land. Market-determined prices, climatic advantage, rapid rise in demand and the producers' access to improved varieties of seeds and subsidized fertilizers have, together, contributed to the fast expansion in high-value vegetables. From 1970 to 1985, the area of tomatoes, one of the more profitable vegetables, tripled in Algeria, doubled in Egypt, Libya and in Sudan, and increased by 60 percent in Morocco and Tunisia. The result of these efforts has been a significant contribution of the vegetables subsector to agricultural GDP in 1980: 10 percent in Algeria; 13 percent in Morocco; 19 percent in Tunisia and 17 percent in Egypt.[3]

Neglect of Livestock Production. In contrast to the rapid technical progress made in the production of vegetables, livestock (cattle, sheep and goats) has been neglected, and has, since the 1960's, been declining in numbers (except in Egypt and Libya). Yet, average growth rate in numbers in 1960-1985, was lower in North Africa than in Sub-Saharan Africa, at 1.1 and 1.6 percent, respectvely. Part of this poor performance is caused by animal death during prolonged droughts and poor management of range and pasture. To meet the fast growing demand for milk and meat, and to provide nomadic people with economic security, livestock production has to grow faster than in the past, through improved breeds, adequate veterinary service cand better use of feed concentrates.

By way of imported breeds, Libya, Morocco, Algeria and Tunisia have improved livestock yields in terms of kilogram of meat per head of cattle and milk per cow. During the period 1970-1985, production of meat and milk increased at the high annual rates of 12 percent in Libya and 7 percent in Tunisia, where animal products contributed about one quarter of gross value of agricultural output.[4] Subsidized feed concentrates and rising demand for livestock products have stimulated this growth. However, livestock products growth and levels of yields are still far below the potential. Towards its realization, a World Bank study (Cleaver, 1982) recommended greater investment in livestock development. The recommendation was based on a subtle analytic reasoning: the economic rate of return computations for livestock component of rural development projects was found to be in the 15-30 percent range.

TABLE 7.2 Yields of Major Food Crops and Livestock in North Africa, 1961 - 1985

Item	Period	Algeria	Egypt	Libya	Morocco	Sudan	Tunisia
Food Crops, Yield Ton/ha.							
All Cereals	1961-1965	0.62	3.31	0.25	0.82	0.82	0.42
	1979-1985	0.66	4.17	0.55	0.87	0.56	0.82
	change %	-6	+26	+120	+6	-41	+95
Wheat	1961-1965	0.64	2.62	0.25	0.85	1.31	0.49
	1979-1985	0.67	3.40	0.70	1.05	1.05	1.04
	change %	+5	+30	+180	+24	-25	+112
Potatoes	1961-1965	6.50	16.30	5.95	9.80	12.05	20.4
	1979-1985	6.60	17.92	6.93	16.62	19.20	12.8
	change %	+2	+8	+16	+39	+59	-37
Tomatoes	1961-1965	10.00	14.90	15.50	16.30	n.a	10.50
	1979-1985	9.75	20.03	13.21	23.50	n.a	19.29
	change %	-2.5	+34	-14.7	+44		+84
Sugar beet	1961-1965	5.70	–	–	20.20	–	20.20
	1979-1985	20.50	27.60	–	38.70	–	32.50
	change %	+263	–	–	+92	–	+61
Livestock							
Meat (beef & veal) Kg/head	1961-1965	117	174	116	118	175	99
	1979-1985	127	144	200	166	165	145
	change %	+8	-17	+72	+41	-8	+46
Kg.milk per cow	1961-1965	640	674	400	500	520	643
	1979-1985	940	673	1,460	629	501	1,182
	change %	+94	-0.1	+265	+258	-4	+84

Source: *Production Yearbook*, Volumes 28, 39 and 41, FAO, Rome

Had the livestock production-sector received a share in total agricultural investment corresponding to its share in total output, the sector's contribution to agricultural growth could have been much higher. According to the Ministries of Agriculture fiscal budgets during the period 1960-1982, livestock sector received between 4 and 11 percent in Tunisia and Egypt, compared to the share of livestock products in agricultural GDP of 28 percent and 30 percent in 1980, respectively.

The indicators of Sudan's livestock sector show an alarming absolute decline in such an important sector which provides a large section of the rural population with food, employment and income. The growth in the numbers of cattle, sheep and goats in 1980 - 1989 has slowed down to one-fourth its average level in 1960-1965. Livestock yields growth, shown in Table 7.2, is lower (and negative) in Sudan than in other North African countries. Furthermore, animal health services are poor. In 1986, there was one veterinarian to 29 thousand livestock units, compared to an average of 2,300 units in the other five countries (*FAO, 1986, p. 61*).

Rapid Mechanization of Agriculture

Of all technical inputs, mechanization of farming operations is the most controversial. The controversy is about such issues as the social cost of labor displacement and the alternative use of scarce capital to generate agricultural growth efficiently. Unfortunately, there is a scarcity of research on these issues. However, Table 7.1 shows a rapid spread of tractors, combine harvesters and threshers in the region, even in the densely populated and capital-scarce Egypt and Morocco, whose wage-dependent landless workers represent nearly one-third of total rural households.

The expansion in mechanization has been sponsored and supported by governments, through generous subsidies and credit supply on concessional terms, mostly in favor of large farmers. With the exception of a small proportion of tractors that are assembled in Egypt, the bulk of combine harvesters, threshers and tractors in North Africa are imported, using up scarce foreign exchange. In some countries, they are part of foreign aid and loans for financing integrated agricultural/rural development projects.

Tractors are useful in certain weather and soil conditions: quick land and seedbed preparation for sowing seeds before the short season of rainfall. Whereas, a plow and two animals take 6-7 days to plough one hectare, a tractor can do it better in only 2 hours, depending on the type of soil. This time factor is an important advantage over the long time taken by animal-powered plows. Tractors also create jobs for skilled workers (drivers and mechanics), and they displace draft animal inputs, thereby contributing to greater production of animals' milk and meat. The use of combine harvesters may be economically desirable in situations where capital is plentiful, labor is scarce, real wage rates are high, and farm size is large enough to warrant full-time, efficient use of harvesters and large tractors.[5] With the exception of Libya, these requisites hardly exist in the other countries.

On examining the intensity of tractor use in relation to wheat yields, we find that Libya has inefficiently overinvested in tractorization. It has used the abundant capital from the plentiful flow of oil revenues to rapidly increase the ratio, from 1,000 ha. per tractor in 1960, up to 77 ha. per tractor in 1985. Although this is the highest ratio in the entire Middle East, it is in sharp contrast to Libya's wheat yields which are among the lowest in North Africa. Considering the fundamental variations in structural, policy, managerial and institutional framework that exist between Libya and the industrialized economies, the Libyan ratio exceeds that in USA and Canada in 1985: 40 ha. and 63 ha. per tractor respectively, compared to 77 ha. in Libya. But the Libyan ratio is sharply contrasted with productivity in terms of average yields of wheat in 1983-85: 2.6 tons per ha. in USA, 1.9 in Canada, in comparison with a very low yield of 0.7 in Libya (FAO *Production Yearbook*, 1985 and 1986). The comparison may be helped by noting that the density of agricultural population on arable land is almost the same in the three countries, i.e. 0.1. But in 1983-1985, average GNP per person in USA was double that in Libya, and it was 72 percent higher in Canada than Libya.

Egypt, being a capital-scarce and heavily populated country, has intensified the use of irrigation pumps more than the use of tractors and threshers. However, the pace of mechanization has, since 1980, been accelerated by the government policy, supported by a substantial financing from the World Bank, the USAID and the Japanese government. The assumption was an increasing scarcity of adult-male laborers in agriculture, owing to their migration to cities and to oil-rich Arab States. With falling real wages as a consequence of migrants' return (during and after the Gulf crisis) and the effect of Egypt's deep recession on rising unemployment, serious doubts have been raised about the implications of the entire program for rural employment. Furthermore, with the removal of farm machinery and fuel subsidies, as part of the economic reform package, capital investment in the purchase of tractors and threshers has become less remunerative, and its private and social opportunity cost benefits in alternative activities have become more valuable (e.g. technologies in augmenting the production of vegetables, fruits, poultry and beef-meat).

Farm Size and Productivity: Have Islamic Principles Affected Land Fragmentation?

The technologies we have just discussed and their complementarity in driving agricultural growth tend to differently benefit farmers by size of holding and by their different capabilities to respond to technological change, and to bear the risk. Whereas there is a general agreement among policy-makers and academics about the social benefit of increasing agricultural production in *all* farm sizes, controversy arises over:

1. Variation in land-productivity levels by size of holdings.
2. The share of different size classes to total food production.

3. The causes of land fragmentation.

The task of this section is to explore these controversial issues, supported by empirical evidence from countries' experience.

TABLE 7.3 The Extent of Small Landholdings and Their Fragmentation in Algeria, Egypt, Libya, Morocco and Tunisia, 1950 - 1982

Country	Year	*Small Holdings as Percent of Total Number*	*Average Area of Small Holdings hectares*	*All sizes of Holdings hectares*	*Average Number of Parcels per Holding in All Size Classes*	*Average Area of Parcels hectares*
Algeria	1973	50.1	1.7	6.2	n.a	n.a
Egypt	1950	78.5	0.7	2.3	2.5	n.a
	1961	84.0	0.6	1.6	2.7	n.a
	1982	90.0	0.6	1.1	2.2	0.9
Libya	1974	n.a	n.a	13.0	3.4	3.8
Morocco	1974	74	1.6	4.9	7.0	0.8
	1962	70	2.0	5.1	n.a	n.a
Tunisia	1961	41	2.3	15.4	4.2	n.a
	1980	44	2.2	14.4	n.a	n.a

Notes: n.a. stands for not available. For Algeria, Morocco and Tunisia, the small holding is under five hectares. For Egypt it is under 5 feddans or 2.1 ha. Number of parcels per holding refers to holdings which have more than one parcel, i.e. a holding in one parcel is not fragmented.

Sources: Results of the censuses of agriculture except Tunisia, 1980 which is from *Enquête Agricole de Base,* and the National Institute of Statistics (INS), Tunis.

Landholding Size and Productivity

The prevalence of small landholdings (or farms)[6] is a major feature of North African agrarian structure. There is no consensus on the criteria used in defining and classifying a holding as small or large. However, the size considered by each country as small is used in Table 7.3 which presents data for five countries that have carried out at least one agricultural census. In our discussion, the emphasis is on the magnitude of small farms, and whether they have constrained or contributed to total food production.

The Extent of Small Holdings. Table 7.3 shows that the percentage of small holdings is increasingly high, particularly in Egypt and Morocco, implying the influence of population pressure on scarce agricultural land. However, the data should be considered country-specific, and therefore, does not permit perfect inter-country comparisons for three reasons. The first is the variation in land productivity, i.e. whether the land is irrigated or rainfed, fertile or non-fertile, etc. The second is the difference among countries in the use of the concept of holding: whereas Algeria included only private holdings in the census, others covered all holdings (private, co-operative and state farms). The third reason is that the size of small holdings is arbitrarily established as a cut-off point and is, therefore, disputable.

Given these limitations, Table 7.3 suggests two tendencies. One is an increasing percentage of small farms in Egypt and Tunisia, combined with a decline in their average area which implies a relative increase in the proportion of poor landholders. The Moroccan opposite direction of change is perhaps due to accelerated implementation of land reform and land consolidation program in *'périmètres d'irrigation'* in the second half of the 1970s. The other tendency is the high ratio between the average area per holding at the aggregate level and that of small holdings, ranging from 7 times as much in Tunisia, to less than twice in Egypt.

But does the prevalence of small-sized farms bring about adverse production effects? In principle, they should not. Empirical evidence indicates that small farms intensify the use of both land and family labor. The results of the 1982 agricultural census in Egypt and Morocco point out that small holdings have a much higher degree of land-use and cropping intensity than large holdings as indicated below by size in percentage.

	Egypt		*Morocco*	
Cropped area	under 2 ha.	over 20 ha.	under 5 ha.	over 20 ha.
per holding	99.2%	55-75%	79%	2-8%

Furthermore, an inverse relationship between output per unit of land and farm size was confirmed by the findings of two penetrating studies on Egypt and Sudan.[7]

Another indicator is output per unit of land by size of farm. In irrigated areas, crop yields and gross value of output per hectare were higher in small farms compared to larger ones of Algeria, Morocco and Tunisia according to the findings of a World Bank study. The study reports "there is evidence in all three countries that small farmers in irrigated areas are more efficient than larger farmers. In rainfed areas, the most efficient farms are medium sized: 15-60 hectares" (Cleaver, 1982, p. 50). The results of a field study conducted in Morocco by ICARDA in 1987-1988 show that the estimated higher net profit per hectare in medium sized than small farms in rain-fed area was due to intensive use of chemical fertilizers and tractors, whereby wheat yields were almost double the level in small farms.[8] Thus, it seems that, after differences in land quality are allowed, the imperfection in the capital market is the most important factor behind productivity variations between small and medium farms.

Labor Use by Farm Size. Laborer-days per hectare tend to decline as farm size increases. This relationship was revealed by hard evidence from Egypt and Sudan (Commander 1987, Tables 4.3 smf 8.4, and Cornia, 1985, Table 2, respectively). In Sudan, the study indicates that man-days per hectare in small farms (under 2 ha.) were on average,three times as much the number in larger farms in the size class of over 15 hectares. The Egyptian study conducted in 1984 reveals that calculated average productivity per labor-hour (physical output in kilograms) for wheat and cotton was higher in small farms below 10 feddans (4.1 ha.) than in farms above that size. The difference is mostly due to an intensive use of family labor on small farms because the surveyed farms were all irrigated, material inputs (fertilizers, seed, insecticide, water pumps and fuel) were accessible to all size farms at a heavily subsidized prices, and output price-level variance across farm sizes was very little at the time of the survey.

Seemingly, these productivity advantages and merits of small over large farms were among the principal factors behind the countries land reform programs, breaking-up large estates, and, in particular, those of absentee owners, for their redistribution, in small units, among poor and landless farmers. Combined with government-technical support to beneficiaries, the fundamental economic aim was the realization of potential output and employment gains from the pre-reform underutilized land and labor. A considerable progress has been made towards attaining this aim. Productivity of agricultural labor has increased between 1970 and 1980 in terms of comparable international dollar, computed by FAO (1986, Table 5.6) as follows:

	Algeria	Egypt	Morocco	Tunisia
1970	292	415	336	449
1980	732	980	841	1,870
Index: 1970 = 100	251	236	250	416

The substantial rise in agricultural labor productivity in Tunisia can be attributed to constant number of agricultural labor force in 1970-1980, while the real value of output tripled in one decade, (increased from International $ 271.9 million in 1970 to $ 1,163.8 million in 1980). In the case of Egypt whose rise in labor productivity value is the lowest among the four countries, the number of working people in agriculture increased by 18 percent between 1970 and 1980.

Food Production and Farm Size. Both census data and village level studies show that small landholders tend to be the primary producers of food crops. They also show the important role played by women and girls in the production of food crops and milk as well as in raising livestock and poultry.

We disaggregated the results of the 1982 census of agriculture for Egypt and Morocco by cropped area in the size of 2 ha. (Egypt) and below 5 ha. (Morocco). It was found that out of the total areas of cereals, legumes and vegetables in Egypt, the share of small holdings was 77, 75 and 43 percent, respectively. With regard to Morocco, the share was 50, 26 and 34 percent, respectively. However, the results of the 1980/81 *enquête agricole* of Tunisia show that only 18 percent of total cereal area and about 48 percent of vegetables area were in the size class of less than 10 ha. Cereals in Tunisia are mostly produced in the medium-size class of 10 - 50 ha., perhaps due to the disproportionate high share of the holders in technical inputs (irrigation water, tractors, threshers, fuel and high-yielding seeds of wheat). For example, the *'enquête'* of 1980 shows that, whereas landholders in the size class of under 10 ha. represented 63 percent of total number and 19 per cent of total area, they received only 13 percent of total high-yielding seeds of hard wheat (durum) which is a staple food. On the other hand, large holders (50 ha. and more each) representing 4 percent of total numbers and 32 percent of total area received 85 percent of this subsidized variety of seeds, but their share in total cereal area was only 37 percent. Radwan and Thomson (1981) reported a similar situation of disproportionate high share of large farmers in Morocco owning over 100 ha.

Fragmentation of Landholdings

The continuing subdivision of all farm sizes and their successive fragmentation have caused concern among rural development policy-makers and analysts. The concern is about efficiency in resource use and possible adverse distributional consequences. But first, it is useful to clarify the difference between the terms 'subdivision' and 'fragmentation'. The former refers to the process of dividing up a single landholding into several farming units, usually through sale and inheritance arrangements. Fragmentation, on the other hand, denotes the *scattered* feature of parcelation, by which the non-contiguous parcels of a single holding are often separated by different distances.

Available information on average number of parcels per holding and average size of parcels is presented in Table 7.3 which shows that the most fragmented holdings are in Morocco and Tunisia. In the former, an area of one parcel is one-tenth of the average area of total holdings. However, the use of averages conceals the extent and distribution of fragmented holdings as well as variation by locality in each country. In Moroccan province, Taounate, 68 percent of total holdings in Karia ba-Mohamed district were below 10 ha. fragmented into 12 tiny plots on average, each about 0.7 ha., while the average number of plots was 7 at the agricultural sector level (Tully: 1990, p. 82). In Egypt, the 1982 census of agriculture shows that 65 percent of total holdings have more than one parcel per holding and that nearly 15 percent of them are fragmented into more than 4 parcels each. Moreover, the size of some parcels is so minute, that on average, each is about one-tenth of a hectare, while the national average was almost one hectare.

Clearly, where fragmentation is extensive, it presents technical and managerial problems which have an economic dimension: constraining effective use of machinery and scarce irrigation water; bringing additional expenses for transport, and for establishing boundaries, fences and small canals; and costs incurred in the supervision of labor across dispersed plots.

The Role of Inheritance in Land Fragmentation

At this point in discussion, one expects the reader to rightly ask: Considering that North African people are predominantly Moslems, are inheritance arrangements responsible for the continual subdivision of farms from one generation to the next? To address oneself to this complex question, we explore the possible causes, breaking them into two sub-groups: structural and institutional factors other than inheritance arrangements; and the interpretation of Islamic principles on inheritance..

Non-Inheritance Factors. Foremost among the structural factors is the fast growing population and their persistent demand for owning and leasing pieces of the scarce agricultural land, as the most secure form of holding wealth and gaining social and political advantages. The average size of landholdings tends to be much higher in less densely populated countries (Libya, Algeria, Sudan) than in Egypt, Morocco and Tunisia. Moreover, with increasing land profitability and the demand for its lease, absentee landowners rent out part or all of their land, resulting in the parcelation of a single ownership holding into several leased, small-sized operational holdings. Likewise, many small landowners tend to expand their holdings by renting-in additional areas in separate plots, resulting in operating two or three distant parcels composing one holding.

Also, there is a tendency among land speculators and international migrants to use their profits and remittances to attract small landowners to sell parcels of their land at inflated prices. Subdivision of land has also been accelerated by rapid urbanization and the construction boom in the 1970s and 1980s, splitting well situated holdings into small plots of farm land, and taking the rest out of production for the construction of houses, factories and roads. Another structural factor influencing the pace of parcelation is longevity. In broad terms, the longer the landowner lives, the lower the pace in splitting up his or her property among heirs, and the longer the cycle in the transfer of land title would be.

Land policy (redistributive land reform and settlement schemes) is also responsible for the small size of farming units and their fragmentation. As indicated earlier in Chapters 4-6, the distributed units -- that the politicians considered adequate at the time of initiating the programs -- proved inadequate after about twenty years. Moreover, most governments have been unable to enforce legal provisions prohibiting the subdivision of holdings below the prescribed minimum size. Even in government-planned settlement schemes as in

Egypt, Libya, and Sudan the allotted areas were smaller than the planned sizes. In part, this is due to political pressure applied on the implementing departments to satisfy more applicants than the number required. For example, in newly reclaimed and irrigated land in Egypt, the settlers' units were allotted in scattered plots, and in 37 percent of the total number of distributed units, the distance between each settler's plots was more than one kilometer (Wazzan: 1974, see Appendix Table 2).

Interpretation of Islamic Principles. Having outlined some structural and institutional factors influencing the subdivision and fragmentation of land, the discussion proceeds to examine the effects of inheritance arrangements which are widely claimed to be *the* cause of parcelation.

Islamic principles of inheritance *'Mirath'* aim at the preservation of the deceased person's estate within his or her family, and guarantee the rights of legitimate inheritors and creditors. The principles also give the estate owner the option to allocate, after death, up to one-third of his or her private property for charity purposes and to selected family members (*wasseyya*). The balance constitutes the compulsory inheritance (*fouroud al-tarikah*) to be subdivided in clearly stipulated shares among entitled heirs, after meeting the funeral expenses and debts owed to creditors. These arrangements are laid in precise terms in the *Qur'ān*, mainly *Surat al-Nisaa*: 11-13 and *al-Baqarah*: 241. Subsequently, they were explained in detail by the Prophet Mohammad in *Hadith*, and by the jurists' interpretation of several principles, in particular those on the subdivision of indebted property and the shares of non-primary inheritors (Abu-Zahra: 1963, p. 25).

Islamic Arrangements for Avoiding Land Subdivision. Two main sets of arrangements are provided by Islamic principles. The first and fundamental is the subdivision of inherited estate in shares *'nasseeb'* i.e. one-half, one-fourth, one-third, one-sixth etc., whereby the male's share is double that of a female heir. The second is *'takharrog'*, by which a heir may withdraw in favor of one or all the remaining inheritors, against compensation to be volitionally agreed upon by the parties concerned. For example, a person owns 20 acres of land and he is inherited by his sister and two brothers, and let us suppose that the sister opts for withdrawal in favour of one brother against compensation payments. The arrangements for the distribution of shares *before* withdrawal are: 8 acres for each brother and 4 acres for the sister. *After* the sister's withdrawal, the shares are: 8 acres for one brother and 12 acres for the other brother (4+8) with whom she agreed to give her share. Clearly, the extent of land subdivision can be further reduced with the extension of this arrangement beyond two inheritors.

As we understand these principles and their interpretations, the proportional rights of the heirs are mandatory, but the terms of their execution are subject to the inheritors' mutual agreement. Once they are identified and their shares applied, the *physical* splitting of inherited area of land is not mandatory. It can be retained and managed as a single production unit on behalf of the heirs by one of them whom they trust, and who is the most experienced in farming. If his (or her)financial capacity permits, he (or she) can pay to other inheritors the value of their corresponding shares. Islamic financing institutions can also lend him the amount of the required compensation. The maintenance of farmland and its protection from deterioration is a virtue in Islamic tradition for the benefit of all concerned *'masāleh, falāh'*.

Islamic Principles Are Not To Blame. After making this attempt to examine the multiple factors which influence land parcelation, what can we conclude? Whereas we cannot underestimate the role of inheritance arrangements, they are neither to be *solely* blamed for the continued subdivision of agricultural land, nor for the limited application of the principle of *'takharrog'* which provides for arrangements to minimize the splitting of inherited land. Nor are they to blame for the inability of governments to enforce existing laws on setting limits below which an ownership cannot be divided and registered. Our discussion suggests that there are also other structural and institutional factors operating in the national and rural economies. These non-inheritance factors contribute, in varying degrees, to the extent of subdivision and fragmentation of agricultural land, and, in turn, to the resulting adverse production and distributional effects.

Redistributive Equity and Rural Poverty

So far we have explored the variations in agricultural growth owing to technological change, with special emphasis on the production aspects of farm size and the multirooted causes of the continual subdivision of farms. The task of this section is to examine the countries' growth performance in relation to land concentration, poverty incidence and investment in human capital.[9] These equity variables, together with agricultural growth, are perceived in this study to be crucial for bringing about a sustainable rural development.

In this inter-country comparison the assessment of rural development performance is, therefore, two-dimensional with distributive equity and agricultural output growth occupying the central concern. Our proposition is that: (1) in an equitable growth path to rural development, this combination, together with increasing earnings from non-agricultural sources reduce rural poverty faster than the sole reliance on agricultural growth; and (2) dynamic growth of agricultural output, though imperative, is not conditional upon the degree of land concentration, i.e. the predominance of large farms is *not* necessarily associated with higher rates of agricultural GDP growth. (See an elaboration of these propositions in Chapter 1, pp. 4-5).

Distribution of Land and Growth Benefits

It is evident from the preceding sections that there has been an unprecedented technical change in agriculture throughout North Africa. Despite output level fluctuations during the last three decades, annual rates of growth were around 3 percent, and in some countries, the rates were even higher at 5-6 percent. This is an achievement, compared to the lower average growth rates in Sub-Saharan Africa and the Middle East (see p.122).

Whereas there is a great deal of truth in the proposition that agricultural growth is essential for rural development, it has to be qualified by emphasizing the length of time that growth takes to significantly reduce poverty, both proportionately and in absolute numbers of the current generation of the rural poor. As argued in Chapter 2, the missing link between growth and poverty reduction is the distribution of growth benefits in favor of poor cultivators and landless workers *via* land reform and other measures of income transfers, targeted to the poor. These benefits enable the poor to develop their capabilities, participate in the rural development process, and to contribute to agricultural GDP growth.

The review of rural development strategies in Chapters 4, 5 and 6 suggests that, over the last four decades, land reform and land settlement schemes have, in different distributive scales, reduced pre-1950 land concentration, provided tenants with security of tenure, and have considerably reduced rental payments in Egypt. There has also been several income-transfer programs by way of food, technical inputs and social services subsidies, and in some countries, the provision of free health, education and purified drinking water to rural people. Nevertheless, with the exception of Algeria and Libya, agrarian reform programs excluded the wage-dependent landless workers from the transfer of property rights. Depending upon the definition used, they represent nearly one-third of agricultural households in Morocco and Egypt, and about one-fifth in Tunisia. Yet the State continues to be the largest single landowner in the six countries.

The degree of inequality of land distribution, together with income transfers and earnings from non-agricultural sources are reflected in the size distribution of income/consumption among rural households. Time-series data from household expenditure surveys are available only for Egypt, Morocco, and Tunisia.[10] We should note that family expenditure is usually lower than income because it excludes savings and purchase of durable goods, hence the less inequality of expenditure than inequality in the distribution of income.

Rural-Urban Income Inequality. Based on the results of these surveys, incomes have always been lower in rural areas than in the cities, and there is considerable inequality within rural areas.[11] Urban per person expenditure in Egypt is 1.5 times that of the rural in 1982, and almost double in Tunisia (1980/81). The average annual expenditure per person in rural Tunisia was 157 dinars and a lower average of 136 dinars per agricultural worker, compared to an average of 332 dinars in urban areas (INS *enquête* results, 1980/81). In Morocco, the ratio of rural-urban inequality falls between that of Egypt and Tunisia: per person expenditure in urban areas was 1.9 times that in rural areas in 1985. This represents a slight reduction in inequality compared to 2.1 times as much in

1970. However, as income is usually more unequally distributed than expenditure, we find from the Ministry of Planning data that at constant 1980 prices, the ratio of average annual income per working person in agriculture to national average was 1:3 in 1985.[12]

Poverty Levels

Inter-country comparison of the incidence of poverty is difficult; the use of different conceptions of poverty lines results in different estimations of who are the poor, and how many they are. Among the several methods of estimation, the head-count ratio (the proportion of the rural population falling below the country-specific poverty line) is widely understandable, particularly by policy-makers.[13] Despite its statistical shortcomings, this method is used in this section.

The variation in country-specific and nutritionally-based poverty lines, defined by different estimators is manifested in the case of Egypt, Morocco and Tunisia. For Egypt, Adams calculated from the 1982 household budget survey total cost of the minimum diet for survival which satisfies minimum calories required by an adult (2,510 calories per day), and by assuming a rural family size of 5 persons, he estimated that 17.8 percent of rural households were poor. Using the data of the same survey, the World Bank calculated the level of income that is barely adequate for the minimum nutritional requirements plus basic non-material needs, valued in monetary terms by the official consumer price index. The result was a higher poverty incidence of 24.2 percent and a rise in the numbers of the poor from Adams' estimate of 4.2 million to 5.1 million. A similar situation existed in Tunisia and Morocco.

In Tunisia, the official estimate of poverty at 14.2 percent in 1980 made by the National Institute of Statistics was based on a low standard of nutritional adequacy identified from the 1980 household budget survey. The low estimate of the rural poor '*défavorisée*' was due partly to the use of low daily calorie requirements of 1,830 *per capita* instead of the norm of 2,200 calories, and partly to weakness in estimating the monetary value of food requirements. By applying the cost of living index to the 1980 poverty line, it was extrapolated for 1985 at 7 percent.

Much higher poverty levels for 1980 and 1985 were calculated by a team of experts from the International Labor Organization (ILO). They followed a different methodology in the analysis of the results of the two family budget surveys. The experts calculated a minimum basic foodstuffs and non-food components (e.g. clothing, housing, transportation, medical care, etc.), which they costed by actual prices for 1980 and 1985. They were, then, able to establish poverty lines per person per year of 109 Tunisian dinars for 1980 and 172 dinars for 1985, compared to the government-defined lines of 60 and 95 dinars, respectively. The result was an estimated rural poverty incidence of 42 percent for 1980 and 31 percent for 1985, in comparison to the government's much lower estimates of 14 percent and 7 percent respectively. The ILO study also identified the poor by occupation; small farmers and farm workers constituting around 37 percent of the country's number of the poor, and the Western districts containing about 70 percent of total rural poor.[14]

With regard to Morocco, there is also a wide divergence of estimates between those worked out by the government and that by the World Bank. Using a minimum annual income per person, the World Bank estimated, in 1975, that 45 percent of the rural population lived in poverty. For 1985, the government's estimate was 17 percent and that of the World Bank was 32 percent. The lower poverty incidence is grossly underestimated because it defined poverty line at only the bottom tenth-percentile of the size distribution of consumption, instead of the twentieth-percentile (the bottom two deciles), recommended by the World Bank. On examining the results of the analysis of the 1984-1985 expenditure survey, we find that the incidence at the bottom two deciles is 35 percent (*Rapport de Synthèse, Résultats*, p. 55).[15] The results of the 1985 survey reveal that poverty is concentrated in the South, where dry farming and nomadism prevail.

Measurement Problems Facing Inter-country Comparisons

Obviously, these differently defined poverty lines and the resulting different estimates of poverty incidence present serious problems in both inter-country comparisons and the formulation of poverty alleviation programs in rural North Africa. They show the lack of a uniform definition of poverty line in a single country, and how such a definition changes from one year to the next, probably for political purposes because the real causes of persisting poverty lie deep in governments' own policy actions.

A major shortcoming of country estimates is in changing the standard of calorie inadequacy which affects estimated poverty levels with great force. Another source of variation is the selection of the price deflator to be used in deriving a comparable poverty line. Moreover, the use of the country's average cost of living index, as the deflator is based on the assumption that a general rise in consumer prices affects the poor and the rich equally. This wrong assumption leads to biases in poverty estimates over time in a single country. The use of price indices specific to rural workers would be more appropriate.

Similarly, the limited availability of data on the size distribution of land in a small number of North African countries, does not permit an adequate measurement of the degree of the association between the variations in poverty levels and land concentration. This limitation inhibits also measuring the relationship between them, and the rates of agricultural growth which are available, timely, and comparable.

Table 7.4   Changes in Agricultural Growth, Land Concentration and Rural
Poverty Incidence in Egypt, Morocco and Tunisia, 1950-1985

	Agric. Growth Average Annual Rates Percent[a]				Land Concentration Gini coefficient and Year[b]	Rural Poverty Incidence Percent and Years of Estimate	
Periods:	1	2	3	4			
Egypt	2.8	2.9	2.9	1.9	0.740 (1950)[c]	56.1 (1950)	
					0.384 (1965)[c]	23.8 (1965)	
					0.432 (1985)[c]	17.8 (1982)	24.2[e] (1982)
Morocco	n.a	4.7	0.8	1.0	0.755 (1974)		45.0[e] (1975)
					0.687 (1982)	17.0 (1985)	32.0e (1985)
Tunisia	n.a	2.0	4.9	4.2	0.622 (1962)	20.0 (1966)	49.0[e] (1966)
					0.604 (1980)[d]	14.2 (1980)	42.0[e] (1980)
						7.0 (1985)	31.0[e] (1985)

Notes: n.a stands for not available.
a. Refers to agriculture GDP. The periods (1, 2, 3, 4) refer to 1950-1960, 1960-1970, 1970-1980 and 1980-1985, respectively.
b. The Gini coefficient is a statistical measure of inequality which ranges from 0 to 1; the larger the index, the greater the degree of inequality.
c. Refers to landownership distribution. The Gini coefficient is calculated by the author from sources given in Table 6.4.
d. Excludes state farms.
e. Estimates made by the World Bank and ILO. See text for different methods followed in estimation.
Sources: Agricultural growth rates are from *World Development Report*, 1983 and 1987.
Land concentration data: Egypt, see note c above. Morocco (1974) is from Table 5.2 and 1982 calculated from the results of the agriculture census. Tunisia (1962) is calculated from the results of *Agriculture Census* and 1980/81 *Enquête de Base*. Rural poverty incidence: Egypt - see Table 6.4 and the text, Morocco and Tunisia - see text "Poverty levels", and notes at the end of the Chapter.

Faced with these data limitations, we are left with the option to describe the data, and to assess the direction, magnitude and pace of change in the three variables presented in Table 7.4 (poverty, growth and land concentration). Another way for understanding the progress made in poverty reduction, which is the core of rural development, is to use the ranking (or ordering) method, after supplementing the few available poverty estimates with the comparable data on the characteristics of poverty given in Table 7.5. They include illiteracy rates, nutritional status, life expectancy and access to safe drinking water. Although the first three characteristics are at the aggregate level, they are useful, considering that a large section of the population is rural. Also they serve as a proxy of governments' commitment to investment in human capital.

No attempt is made to construct a single synthetic index of this quantitative information. It would conceal rather than reveal the importance of each component. Such a technique requires weighting each of the rural development indicators which is "undesirable, unnecessary and misleading". This conclusion was reached by two seasoned scholars, Hicks and Streeten (1981: pp. 387-90), on the basis of their review of work using this method. Nevertheless, the United Nations (UNDP) ventured in 1990 to construct a human development index that composes three key components: income, health, and educational attainment.[16]

Relationships Between Growth, Land Concentration and Poverty

Quantitative measurement of rural development indicators are available for Egypt, Morocco and Tunisia. Although their measurements have already been presented separately in the course of discussion, they are assembled in Table 7.4. The purpose is to assess whether all or two of them are closely related or not. For this purpose, we shall also benefit from the results of other analysis using a larger sample of developing countries which includes Egypt and Morocco. Let us begin with the annual rates of agricultural GDP growth (Y hereafter) and the Gini coefficient of inequality in the size distribution of land (G hereafter).

The Relationship Between Growth and Land Concentration. The data on these two variables for two or more points in time tend to show a weak association, i.e. having no clear observed linear relation. This can be seen from individual country data. Egypt's rates of Y were nearly the same in both periods of 1950-1960 and 1960-1970, despite the sharp reduction in G following the land reforms of 1952 and 1961. However, a different direction of change is found in Tunisia; a considerable rise in Y in the 1970s and 1980s, while G has slightly fallen after the completion of land redistribution program by 1970. Its low Y, in the 1960s, is due to a combination of unfavorable weather effects (droughts of 1962 and 1967), and the inefficient management of production co-operatives -- which were dissolved in 1969 after being rejected by private landowners.

The sharp fall in Moroccan Y, from 4.7 percent in the 1960s to 1.0 percent in the first half of the 1980s, can not be associated with the slight reduction in G.

resulting from a very modest land reform that redistributed only 4 percent of agricultural land. In searching for an explanation, due consideration should be given to variables other than land concentration. They, especially, include weather, disincentive arising from distorted pricing policy, and the neglect of the large traditional sector of rainfed agriculture. Post-1985 agricultural trade liberalization, including price policy reform, and the intensive technical change have, in varying degrees, induced a fast growth of Y, reaching 6.4 percent in 1985-1990 (see Tables 5.3 and 7.1).

This small relationship between Y and G was also found by the author from the results of his regression analysis of agricultural GDP growth (Y) on land concentration (G) in 1973 -1983 (Y as dependent variable) for a larger sample of 20 developing countries, including Egypt and Morocco. The estimation results are not statistically significant, and the coefficient, being very low at 0.04, indicates that only 4 percent of the variation in agricultural growth rates is 'explained' by the degree of inequality of the size distribution of land (El-Ghonemy, 1990, p. 175 and Appendix A).

The Relationship Between Land Concentration and Poverty. Despite the limitations of poverty estimates (P), noted earlier, there do appear to be a positive relationship between P and G. Irrespective of its different estimations, poverty incidence moved downward, together with G in the three countries, as suggested by the data in Table 7.4. This positive association is greater in the case of Egypt whose G fell by 92 percent, and P was reduced much faster by 136 percent from 1951 to 1965. While G was slowly rising in 1965 - 1982, poverty incidence also rose slightly from 23.8 percent to 24.2 percent, the poverty level estimated by the World Bank.

A similar direction of change in G and P relationship took place in both Morocco and Tunisia, in spite of the wide range of P estimates. Clearly, the reduction in P *cannot* be solely explained by the small reduction in G (9% in Morocco and 3% in Tunisia), though its influence is likely to be important. There are several interacting factors which have probably influenced poverty reduction, notably the dynamics of the labor market, and the considerable amounts of remittance receipts in rural areas earned by migrant workers in both countries. Important also is the considerable rise in Tunisia's real income per working person in agriculture noted earlier (p. 130).

The Pace of Poverty Reduction. A recently conducted multiple regression and cross-sectional analysis of the values of P, G and Y in 21 developing countries, including Egypt and Morocco, estimated the pace of poverty reduction in relation to agricultural growth and land concentration (Tyler, El-Ghonemy and Couvreur, forthcoming). The analysis indicates a positive, strong correlation between the variations in P and G (0.7), and is highly significant. The estimation indicates that with one percent decrease in the Gini coefficient of land concentration (G), it is expected that poverty incidence would fall by 1.6 percent.

These results imply that at a 3 percent annual growth of agricultural GDP per head, it would take 60 years to reduce poverty level by only half (50 percent), while the same proportion of reduction could be attained much faster by a one-third decrease in G. This estimation fits well the case of Egypt. Between 1951 and 1965, Egypt implemented two land reforms, and at 3 percent annual agricultural GDP growth, poverty incidence was reduced to nearly half its level in 1951 (from 56% to 24%). This sizeable reduction occurred in only 14 years, instead of the estimated period of 60 years (till the year 2010) *if* Egypt relied solely on agricultural growth, without substantially decreasing land concentration.

Reducing the numbers of the rural poor does not only depend on land reform, but also on the country's policy to slow the rate of population growth and the related fertility rate.[17] Abated population growth and small family size are conducive to poverty alleviation, through a firm government commitment to slowing population growth (family planning), which is unopposed by Moslem leadership, notably *al-Azhar*. Of the four countries for which poverty estimates are available, there is a close association between poverty incidence and rates of population growth and fertility.[18] Sudan and Morocco have the highest levels of both poverty and population growth-cum-fertility. In such a situation, poverty reduction may not be possible unless slower population growth can be achieved soon. On the other hand, Tunisia has the lowest levels, and it is the first country in North Africa to have, since 1963, embarked on an official but serious program for family planning. The manufacture and sale of contraceptives are heavily subsidized by the government, abortion and sterilization are legalized, and polygamy is abolished. All these policy measures for slowing population growth have been integrated in rural development programs and, since 1982, resources have been concentrated in North-Western and Southern provinces where rural poverty is high.[19]

Investing in Human Resources

To overcome the absence of data on the incidence of rural poverty in Algeria and Libya, and the limitations of estimates in the other four countries, the quantitative and comparable indicators of public investment in human resources are assembled, and presented in Table 7.5. The data include the United Nations' human development index (HDI) and public expenditure on health and education. Clearly, these indicators overlap, and influence each other. Progress in nutrition, literacy and the provision of safe drinking water are reflected in life expectancy. Illiteracy rate and life expectancy are two components of HDI which also includes real income per person. Likewise, the share of health and educational services in total public expenditure is manifested in illiteracy rate, access to purified drinking water and life expectancy.

Table 7.5 Changes in Nutrition, Health and Illiteracy Indicators and Share of Health and Education in Public Expenditure in North Africa, 1960 - 1985

	Public Expend. on Health & Education % of Total 1972, 1982	Average Annual Growth Rate of Calorie Supply 1969-71 to 1979-81 %	Life Expectancy Increase 1960-1985 (percent)	Rural Popn. with Access to Safe Drinking Water 1985-87 (percent)	Percentage Illiteracy reduction 1960-85 (percent)	Human Development Index
	1	2	3	4	5	6
Algeria n.a		3.6 (2)	33 (3)	55 (3)	41 (2)	0.609 (3)
Egypt 13 12		2.2 (3)	35 (1)	56 (2)	21 (4)	0.501 (4)
Libya n.a 24		4.7 (1)	34 (2)	90 (1)	49 (1)	0.719 (1)
Morocco 24 19		0.7 (6)	31 (4)	27 (5)	15 (6)	0.489 (5)
Sudan 18 15		1.0 (5)	18 (6)	10 (6)	20 (5)	0.255 (6)
Tunisia 38 29		2.0 (4)	31 (4)	31 (4)	22 (3)	0.657 (2)
Average Middle Income Countries 20 16		- -	27	50	32	

Notes: n.a. stands for not available. Years are selected to be around or close to years of poverty estimates given in Table 7.4. Numbers in brackets are the ranking order of countries, whereby (1) refers to the best performer and (6) the worst. Life expectency is the number of years an infant would live under prevailing conditions. Illiteracy is the percentage of adults aged 15 and over who cannot read or write. Human Development Index is defined in note 16 at the end of the chapter. These three indicators are at the national level.

Sources: Column 1 is from *World Development Report 1985*, World Bank, Oxford University Press, except Egypt, the first year refers to 1977, and is from IMF *Government Finance Statistics*, Vol-VII, and Libya, 1982, is from the Ministry of Planning.
Column 2 is from *The Fifth World Food Survey*, FAO, Rome, 1985, Appendix A.
Columns 3 and 5 are calculated from *World Development Report* 1978 , 1987, Indicators.
Column 4 is an average calculated from *The State of the World's Children* 1986, UNICEF, Oxford University Press, and *Human Development Report* 1990, UNDP, Oxford University Press. Column 6 is also from *Human Development Report*, 1990.

Although data on nutrition, life expectancy, and illiteracy are available up to 1990, the time-frame of our study is limited to 1960-1985; a period closer to the years of land concentration and poverty estimations given in Table 7.4. The ranking order of the countries, relative to each other for individual indicators, puts Libya at the top, followed by Algeria, Egypt and Tunisia. It also ranks Morocco and Sudan at the bottom. In four out of the five indicators, Libya has the highest ranking (1) and is ranked second in only one indicator (life expectancy). For the composite HDI, Libya stands at the top of the six countries, followed by Tunisia and Algeria. This is probably due to the influence of Libya's high GNP per head, being an indicator of the national income used in the HDI. The same ranking order exists in the share of public expenditure on health as a percentage of GNP (1982-1986).[20]

In the absence of disaggregated data by rural and urban sectors, it is difficult to judge the disparity between national and rural quality of life indicators. One could venture to say that, in all North African countries, illiteracy (particularly among women) is much higher in rural than urban areas. Available information from UNESCO and population census reveals that in rural areas of Egypt, Morocco, Sudan and Tunisia, the rates of illiteracy and school girls' drop-outs are double those in cities. The implications of this lost human capital formation among more than half the rural population are serious. There are positive correlations between child malnutrition, infant mortality and illiterate mothers.[21]

Assuming that cities have 100 percent access to purified drinking water, in 1985-1987, we find that the narrowest gap between rural and urban areas exists in Libya, Egypt and Algeria. Again, the widest gap is in Sudan followed by Morocco. From 1960 to 1985, the notable achievement by all countries (except Sudan) is raising of the average life expectancy at rates higher than the weighted average of the 58 middle-income countries and that of sub-Saharan Africa. Likewise, progress made in nutrition level is notable; only Morocco and Sudan are below the average of developing countries. We recall that both countries failed to satisfy the minimum nutritional requirements for the rural poor, estimated by the World Bank, in the early 1980s, at 32-45 percent of rural population in Morocco and 80-85 percent in Sudan.

In reality, with statistics and without using statistical jargon in counting the poor, their lives, nutrition and educational status are intimately connected. Investing in rural people's health and education yields high social benefits and economic return to both individuals and the North African economies.

Notes

1. The dominant public sector in total investment is in contrast to the low share of private sector estimated by the Investment Policy Committee of the Egyptian Parliament *Maglis al-Shoura'* at 4.1 percent in 1959-1966 and 12.5 percent between 1974 and 1981, cited in Hassan Khedr, 'Public Expenditure and Agriculture Taxation', 1989, Table 4.3, mimeographed.

2. This comparison was made in 'Socio-Economic Indicators for Monitoring and Evaluation of Agrarian Reform and Rural Development in Egypt', a study prepared by the Ministry of Agriculture, mimeographed in Arabic.

3. For Algeria, Morocco and Tunisia, see the World Bank study, *The Agricultural Development Experience of Algeria, Morocco and Tunisia,* A World Bank Staff Working paper, No. 552, 1982, p. 14. For Egypt the contribution of vegetables to gross value of agricultural output is calculated from Table 1A of *al-Iqtisaād al-Zirā'i* (Agricultural Economic Bulletin), 1983, in Arabic.

, 4. These rates of livestock products' growth are calculated from *Country Tables* 1991, FAO, and the contribution of livestock production to agricultural GDP in Tunisia is from Radwan *et al, Tunisia: Rural Labor & Structural Transformation,* Routledge, 1991, p. 36.

5. By efficiency is meant high return on capital and scarce manpower (skilled management and labor), and is measured in terms of output per laborer-hour.

6. Agricultural landholding is an operational, economic unit managed by a single holder, irrespective of his or her legal tenure status. It is not an ownership unit, only if managed by its owner, is classified as owner-operated holding. If two or more persons jointly operating the holding are members of the same household, only one of them is considered a landholder. For further classification of a holding, see the section 'Holdings and Holders' in *1970 World Census of Agriculture, Analysis and International Comparison of the Results,* FAO, Rome, 1988 pp. 16-23.

7. For the relationship in Egypt, see Simon Commander, 1987, Table 8.1. For Sudan, see Cornia, "Farm Size, Land Yields and the Agricultural Production Function: An Analysis for Fifteen Developing Countries', *World Development* Vol. 13, No. 4, Tables 2 and 3, 1985.

8. This study was conducted during 1987-1988 in 210 farms (68% below 10 ha., 28% are 10-15 ha. and 4% are in the size class of over 50 ha.) situated in Karia ba-Mohamed district, 65 kilometers North of the city of Fez. Wheat area represented 55% of total cultivated area and the rest grew beans, chickpeas, forage and sunflower. The results of the survey show that average yield of traditional small farms (below 10 ha. each) was 1.1 tons per hectare, compared to 3.5 ton/ha. in mechanized medium farms. Net profit per hectar (after deducting costs of inputs, labor and machinery totalling 3,170 dirhams), was 3,505 dirhams, compared to 1,120 dirhams in small farms (Dennis Tulley, ed., 1990, pp. 81-101).

9. Investment in human capital is viewed here as the provision of adequate opportunities to develop income-yielding abilities in the expectation that future benefits will exceed the costs incurred. By abilities, we mean enhanced capacity of rural people to better perform functions in life, and to have a wide choice of opportunities. This enhancement results from improved health, nutrition and education (including skills) which are considered created capital assets, yielding income over a person's lifespan. For a detailed discussion of this concept and its application, see Schultz and Ram (1979), Schultz (1981), and Arne Bigsten (1983), particularly his section 'The human capital school and its critics' pp. 9-14.

10. These sample surveys, like those household surveys in many other countries, have limitations: neither the very rich nor the very poor households are adequately represented; consumption of self-produced food crops are usually underestimated; and expenditure on social activities tends to be under-reported. Nevertheless, their results are useful in estimating the incidence of undernutrition and rural poverty.

11. These rural-urban income ratios based on averages can be misleading. First, rural incomes include not only the meagre earnings of poor peasants and landless workers, but also the considerably larger incomes of landlords, medium farmers and large-scale traders operating in rural areas. Second, rural income may be partly in cash and partly in kind, which is not easily quantifiable. Third, average urban income is usually underestimated because some rural people generate their incomes in urban areas activities, but they count in household surveys as rural income.

12. Using data on *per capita* total GDP and agriculture GDP *per capita* rural population for the year 1985 (the survey year), we find that at constant prices (1980) the rural-urban ratio was 3,948/1,173 dirham or 1:3 in favor of urban areas (*Annuaire Statistique du Maroc*, 1989 and *Compte et Agrégates de la Nation*, 1980-85). This is a much higher inequality than 1:1.9 calculated by the survey on per person consumption. In 1970-1971 survey, average expenditure *per household* in urban areas was 8,436 dirhams, compared to 3,938 dirhams in rural areas (at current prices).

13. For monitoring progress made in poverty alleviation and for targeting resources more effectively, rural development planners need both an unchanged criteria of defining poverty line and an unambiguous classification of the simple label "rural poor" by region, occupation, land tenure status and by gender.

14. See Radwan *et al*, 1991, Table 4.9.

15. These results are taken from '*Consommation et Dépenses des Ménages*, 1984-1985, Ministère du Plan, Vol. 1, Rabat.

16. Human Development Index is a composite and weighted index, ranging from 0 to 1, the higher the index the higher the human development achievement in a given country. See *Human Development Report* 1991, technical note 1, pp. 88-89.

17. Fertility rate is the average number of children that would be born alive by a woman during her lifetime. According to the *United Nations Yearbook for National Statistics*, total fertility rate in the four countries, for which rural poverty estimates are available, is as follows:

	Egypt	Morocco	Sudan	Tunisia	Developing Countries
1975	5.2	7.1	7.0	6.2	6.2
1985	4.7	4.9	6.6	4.6	4.2

18. Population factor is implied in poverty estimates which are based on the definition of a minimum *per capita* income/expenditure/food consumption and daily calorie requirements for survival. These are used in establishing per person or household poverty lines for rural and urban population.

19. See: *Survey of Laws on Fertility Control*, UN Fund for Population Activities (UNFPA), New York, 1979; Tunisia 4th, 5th and 6th *Plan de Développement Economique et Social*, 1973, 1977 and 1981; *Program d'activities*, 1977-1981, Office National du Plan Familial et de la Population, Tunis, 1977; and '*al-Malâmeh al-Jehaweya lel Mokhatat al-Sadis*' (The Regional Characteristics of the Sixth Development Plan), Commisiorat Général de Développement, Tunis, April 1982, in Arabic.

20. In 1984 - 1987, the share in descending order was: Libya (3.2%), Tunisia (2.2%), Algeria (2.0%), Egypt (1.1%), Morocco (0.9%) and Sudan (0.2%). The average in Sub-Saharan Africa was (1.0%).

21. In their cross-country study in 95 developing countries, FAO and WHO found a simple positive correlation of 0.4 hetween illiteracy rate among women and undernutrition of children (1-2 years). Infant mortality risk is higher and its correlation coefficient was estimated at 0.5 (positive). See *"The Fifth World Food Survey"*, FAO, 1985, Table 4.1.

8

Challenges and Dilemmas

It is difficult for the author of a study like ours, examining six countries' rural development experience over a long span of time, to decide what to say in a final chapter after attempting to make each chapter comprehensive. Perhaps an easy way to conclude is to synthesize the findings which are of interest to policy-makers, development analysts and students of rural development. Granted that this approach is convenient, it is, nevertheless, unimaginative. It denies the reader a discerning vision of challenging long-term development issues, towards the year 2000 and beyond into the next century. For the benefit of the reader, it is perhaps better to perceive the countries' past experience as a set of responses to a series of challenges, and to view present development problems and future policy issues in terms of difficult choices to be made that present policy-makers with dilemmas in rural development.

This suggested approach is reflected in the structure of this chapter. It consists of two major sections. The first examines how politicians and planners have, since the 1950s, responded to the challenges presented in tackling rural underdevelopment problems. This section also explores how the rural people have responded to government policies and its huge bureaucracy, and to the challenge of escaping the risk of poverty and economic insecurity. The second section discusses the dilemmas facing policy-makers in finding satisfactory solutions to short- and long-term rural-agricultural development problems, and in particular those generated by the 1980s' economic reforms.

Response to Past Challenges

Since political independence, North African governments have faced common challenges but reacted differently in tackling the inherited rural development problems. Influenced by varied endowments of natural resources and ideological preferences, the chosen paths created developmental problems which, in turn, have brought about a new set of challenges.

Foremost among these challenges, was how to redress the perpetual imbalanced development within the rural-agricultural sector and between it and the urban-industrial sector, and the adjustment of policy and investment in favor of the former. We realize, however, that 'imbalanced' in contrast to 'balanced' development, is not an unambiguous term. It is surrounded by contention and subject to different interpretations and measurements.[1] As repeatedly stressed in the preceding chapters, we are not concerned with agricultural growth and high yield on capital investment *per se*, but more with what happens to the equitable distribution of their benefits and to the incidence of poverty and malnutrition within rural areas. Empirical evidence presented in the course of country-experience review, indicates a distinctive dualism in agricultural development, and the imbalance in the shares of agriculture, health and education in public spending. This imbalance is not solely dictated by economic considerations. Rather, it is embedded in the socio-political framework of policy design. Although, we are aware that 'agriculture' and 'rural' are not synonymous, there are no serious errors in using the phrase 'rural-agricultural sector' *versus* 'urban-industrial sector'.

The Response of Policy-Makers

At the time of independence and Egypt's revolution led by Nasser in 1952, the countries' new leaders inherited rural-agricultural sectors shaped by vested interests of colonial administration and native landlords in coalition with rich city merchants. To a considerable extent, these vested interests conflicted with those of the masses of rural people who were appallingly deprived of basic social services. Nevertheless, there were already fundamental technological changes in agriculture, especially irrigation combined with institutional credit. Importantly, by 1950 North African agriculture had been closely linked with the world market. But the pattern of commercialization of agriculture created a deeply rooted imbalanced rural development: a duality of agrarian structure having distinctive features of land concentration, on the one hand and landlessness, insecure tenancy and chronic indebtedness, on the other. Another feature was the wide gulf in productivity and income created between the vast subsector of traditional rain-fed agriculture and the small commercial and often irrigated subsector. Hence malnutrition, poverty, illiteracy and gross inequalities of rewards and opportunities prevailed in rural North Africa.

Confronted with these challenging problems, the countries' new leadership differed in the design of rural development strategy but were united by one belief. They believed that the initial problems of underdevelopment could only be solved by a system of state tight control with emphasis on planning and import- substituting industrialization, and a government that undertakes the maximum of functions, reducing the economic freedom of the private sector to a minimum. Prominent policy measures were land reforms, nationalization of foreign enterprises and, with the exception of Morocco and Sudan, the limitation of big landholders' power. The aim was to alleviate injustice, oppression and poverty, through the removal of institutional barriers to rural development.

Clearly, in this broad assessment of stategy choice, policy similarities are exaggerated. Also, some real differences in style, content and pace of change are blurred. In the text, (chapters 4, 5 and 6) such variations are pointed out. Moreover, different distributional and production effects of contrasting strategies were examined countrywise and by rural localities. The purpose was to alert the reader to the ill consequences of an oversimplified approach to rural development policy.

It emerges from our discussion of policy choice that, despite a wide variation in the scope of reforming the inherited agrarian systems, the poor *fellaheen's* long-awaited expectations have been partially fulfilled. However, on balance and in an historical perspective, the two decades 1952-1972 can be called the *fellaheen decades*. Yet, many things went wrong. Country case studies suggest several flaws in both the design and implementation of rural development policies and programs. In order to avoid repetition, we make a few concluding remarks on common defects.

Perpetual Imbalance and Polarization. Dualism and imbalanced rural development have been perpetuated, through incomplete, distorted institutional reforms, the pattern of agricultural growth and misallocation of resources between irrigated and rain-fed sectors, and between food and non-food crops, to the detriment of small farmers. Where ceilings on private land ownership were fixed, they were too high to permit greater access to land by the masses of landless poor farmers (except in Libya). This paradox occurred while there was no shortage of cultivable public-land to distribute. The State, in addition to its ownership of a considerable area of arable land, has retained 10-15 percent of the expropriated land area in Egypt, Morocco and Tunisia for establishing inefficiently managed large state farms.

At the same time, polarization of the size distribution of land and incomes has widened, due partly to persistent demand for land, and partly to governments' own policy of distributing very small units to beneficiaries. Another paradox was in the issue of laws fixing minimum wages for agricultural workers which proved to be unenforceable, particularly in slack seasons when the supply exceeds demand. In retrospective, it seems that many of these policies were used by politicians as symbolic justice, and to pacify the discontented rural poor without harming the interests of large and medium farmers. With these flaws, the momentum for a complete land reform[2], created by the unique political environment in the 1950s and 1960s, was dissipated.

Well meaning policies were formulated and laws issued on the assumption that bureaucracy is capable of delivering the benefits to the *fellaheen*. Alas, in many cases the effectiveness of beneficial programs has been constrained by the administrative manner of their execution. First, cumbersome bureaucracy in implementation has been compounded by a concentration of decision-making at the top. Second, responsibilities for rural development programs have been sparsed over several government departments, having vaguely defined functions, which often overlap without clear accountability. Even the responsibility for the important subject of nutrition is divided between the ministries of agriculture, health and education. Third, the lack of incentives payment and residence allowances for civil servants posted in rural areas has tended to tempt some of them into corruption, principally those working in tax-collection, state farms, and management of village co-operatives. Fourth, passing minor decisions on to the top by qualified field personnel with regard to urgently needed actions has resulted in reduced developmental impact, and in an unnecessary loss of output. In short, the limited scope of redistributive programs and the manner in which they were administered have failed to redress the imbalance within rural areas and between rural and urban sectors.

An Estimation of Imbalanced Development. Developmental imbalance has, since the 1960s, been built into national plans. Invariably, an overriding importance has been given to total output growth as *the* critical objective of development (and the criterion for achievement), and not what happens to the basic amenities of life (health and education), poverty, food insecurity and inequality in income distribution. Hence, there are distorted monitoring and statistical systems, and inadequate targets and data on progress made in rural development, particularly the changes in the incidence of poverty and malnutrition.

Obsessed with the primacy of rapid urbanization and import-substituting industrialization, governments intervened extensively in the agricultural products market, distorted prices, and monopolized the supply of major inputs. In this way, farmers, especially small producers, had, until the mid-1980s, been heavily and indirectly taxed (except in Libya). Added to price intervention, public capital and current expenditures have yielded an imbalanced development with a substantially administered transfer of surpluses from rural-agricultural sector to support other sectors. Governments devoted a disproportionately large share to support the internationally uncompetetive and indiscriminately protected manufacturing industry. Extracted rural surpluses were also used to subsidize urban consumption where bureaucracy is concentrated, resulting in greater rural-urban income inequality. Whereas this imbalance-generating policy was deliberately designed, its adverse consequences were probably unintended. It has been carried out too far at the expense of agriculture and investment in rural infrastructure. This deliberate transfer of income from rural-agricultural sector to urban-industrial sector has occurred despite the convincing findings of IFAD in Sudan and the results of a World Bank study in Algeria, Morocco and Tunisia. By evidence, they established that investment in the former is more efficient than the latter in terms of incremental capital-output ratio (IFAD, 1988 and Cleaver: 1982, p. 18).

The actual experience of North African countries shows that the absolute priority given to industry has neither adequately absorbed a sufficient number of the labor force, nor has it increased correspondingly the share of manufacturing industry in total gross domestic product and exports. Despite increased rural to urban migration, rural-agricultural labor force relative to industry has remained quite large (except in Libya and Tunisia). By denying the agricultural sector its capital needs, output growth either stagnated, or grew slowly which, in turn, has held back national income growth (Chapter 3). Using Balla index, we estimated the degree of imbalance between the rural-agricultural sector and the growth performance of total GDP. For an easy comparison, we also added to the data in Table 8.1, the proportionate labor absorption in agriculture and industry, as well as the share of agriculture in aggregate investment. As explained in the table's notes, it was not possible to include Libya.

The estimated index of imbalance presented in Table 8.1 puts Tunisia at the top and Morocco, followed by Sudan, at the bottom, i.e. the latter two having the highest degree of imbalanced development. With a zero deviation between agricultural and total output growth rates, the index for Tunisia suggests an almost perfect balance. By absorbing labor at an annual rate of 3.7 times as much as agriculture, industry reduced the pressure of rural people on land, and kept the numbers of agricultural labor nearly constant, and, thereby, improved the income position of those who remained on the land. These cumulative forces, combined with positive control of population growth and devoting adequate resources to agriculture, have resulted in a rapid rise in the real income (and nutrition) per working person in agriculture (Chapter 7, p. 130). The opposite took place in Morocco and Sudan, whereby resources devoted to industry were disproportionately high to the disadvantage of output, investment and employment in the rural-agricultural sector. We may recall from Chapter 7 that the characteristics of malnutrition and rural poverty in Sudan and Morocco are consistent with the degree of imbalance in development suggested by Table 8.1. To say that there were not enough resources to invest in agriculture, notably in irrigation, and in the health and education of rural people, while spending about one-quarter of government expenditure on the purchase of arms and defence, is to say that the policy-makers do not really want rural development.

The Response of Rural People

Despite the proclaimed pro-*fellaheen's* rural welfare policy, the masses of low-income rural people had no say in, or control of the design of rural development strategy and programs, and they are not to blame for the resulting imbalanced development. Policies were made by a small group of army officers after *coups d'état* (Egypt in 1952, Libya in 1969 and numerous *coups* in Sudan) and a few leaders of independence movement (Algeria, Morocco, Tunisia and

Sudan). It is true that there were scattered, locally unorganized agrarian unrest in response to pre-reform exploitative production relations. But there was no organized peasant movement leading to radical rural change as happened, for example, in Mexico in 1911, China in the 1940s and Kerala, India in 1959).

TABLE 8.1 Indicators of Imbalanced Development between Rural-Agricultural and Industrial Sectors in North Africa,[a] 1980-1985

Indicator	*Algeria*	*Egypt*	*Morocco*	*Sudan*	*Tunisia*
Imbalance Index[b]					
1. GDP Annual Growth rate %	4.9	5.2	3.0	0.3[c]	4.1
2. Agric. GDP Growth rate %	2.1	1.9	1.0	0.4[c]	4.2
The Deviation $(2 - 1)^2$	7.84	3.3	4.0	0.01	0.1
3. Share of Agric. in Total GDP %	8	22	18	34	17
4. Index	1.61	1.63	2.83	1.94	0
Ranking Order	(2)	(3)	(5)	(4)	(1)
Agric. and Industry shares					
in Total Labor Force average					
5. Agriculture	31	46	46	71	35
6. Industry	27	20	25	8	36
Average Annual Growth of Labor %					
7. In agriculture	0.9	1.9	2.3	1.9	0.3
8. In Industry	0.9	0.9	0.8	0.2	1.1
Ranking order[d]	(2)	(3)	(4)	(5)	(1)
9. *Share of Agric. in Total investment %*	11	9	7	1	13

Notes: a. Libya is excluded from both the imbalance index (because of incomplete data on GDP) and labor absorption (because of the dominant petroleum industry). b. The index is a measurement of the degree of intersectoral imblance. It equals the square root of the deviation multiplied by the share of agric. in total GDP, divided by the average annual growth of GDP. The higher the index, the greater is the deviation, and the higher the degree of imbalance. c. Refers to 1980 - 1986. d. according to the difference in the rates of labor absorption (row 8 - row 7).

Sources: Rows 1 and 2 are from *World Development Report* (WDR) 1987, 1988, World Bank, Oxford University Press, Development Indicators. Rows 5, 6, 7 and 8 are based on data in WDR (development indicators), *ibid,* and *Country Tables: Basic Data on the Agricultural Sector,* 1991, FAO, Rome. Row 9 is taken from Table 7.1.

The methodology used in calculating the imbalance Index is from: A.S. Balla, *Uneven Development in the Third World,* Macmillan Press, London, 1992, p. 28.

Encouraged by the success of initial agrarian reforms, governments extended their power and authority to the rest of the agricultural sector. Backed by an increasing sizeable bureaucracy, the extensive intervention has been implemented at village level by the imposed forms of co-operatives. These over-staffed institutions were entrusted with a wide range of tasks: implementation of land reform; allocation of land among crops; supply of subsidized means of production; and the procurement of crops at government-administered low prices. The response of rural people was a mixture of gratitude for gaining secured rights in land combined with receiving inputs at low price, and discontent about the tight state control of their economic transactions. In some cases, farmers were appalled by the contradiction in morality between the corruption among those officials who claimed to transmit the new idealism of post-colonial reforms, on the one hand, and the new set of values to be practised after the abolition of past semi-feudalistic exploitation, on the other.

Resisting Tight Government Control. From the author's field visits in the 1960s and 1970s, it was evident that whereas small farmers were discontented, they could do nothing but obey the officially established rules in co-operatives' management, which exercized local monopoly power. These institutions were used by policy-makers as an employment outlet for incompetent graduates and a tool to achieve the pre-determined end of serving political motives and accelerating the accumulation of rural surpluses; an end in which small farmers had no say and over which they yield no power.

Small farmers were powerless because they became dependent on what government officials decide and how co-operatives function. They had no grass-roots level voluntary organizations of their choice to lobby for reforming these arrangements, particularly those related to pricing. Trade unions of rural workers are either banned or are government-controlled, and their right to strike is prohibited. This action on the part of governments contradict their commitment at the 1979 World Conference on Agrarian Reform and Rural Development. At that conference, North African governments adopted a Declaration of Principles and a Program of Action, in which they agreed to widespread sharing of political power, and to ratify and enforce the ILO Conventions Nos. 87 and 141 on freedom of association and the rights of rural working people to organize their own associations for improving their social and economic conditions. Only Tunisia and Egypt ratified these Conventions.

The problem, I suppose, seems to lie chiefly in a real dichotomy between the *fellaheen*'s perceptions of what they want and the operative ideology and faulty priorities of politicians at the top. However, with mounting problems of falling productivity and their silent resistance, Tunisian and Algerian farmers, succeeded in influencing politicians to abolish collectives (production co-operatives), and to replace them with regular service and marketing co-operatives, the membership of which is voluntary.

Yet, empirical findings reveal that in rural-agricultural co-operatives (except in agrarian reform areas), the poor cultivators surveyed throughout North Africa, do not benefit -- in most cases -- from their services: the chairman and membership of managing committees continue to be the same influential big and middle-sized landowners; the balance sheets and auditor reports are not discussed at the annual meetings as they are incomprehensible; and government-appointed managers are accountable not to local farmers, but to central ministries.[3]

Saving and Investing. Despite this chain of political and institutional constraints, the rural people -- as producers and consumers -- have met the challenge.   Agrarian reform beneficiaries have, invariably, increased land productivity, principally food. Given that about 60 - 70 percent of their income is usually spent on food, the balance of their outlay is spent on the purchase of fertilizers and non-food items (clothes, footware, furniture, fuel, transport, etc.) which are *domestically* produced. The higher their incomes, the higher is the purchase of these products, and the greater is the employment of labor in the corresponding industries and services in the domestic economy.   Empirical evidence (Chapters 4 - 7) show that poor cultivators invest savings and remittances in the purchase of income-yielding assets (e.g. livestock, irrigation pumps, and small farm machinery). Had they been left with all their surpluses without being indirectly taxed through distorted pricing and moral irregularity of local officials, their contribution to increasing the productive capacity of their countries' rural economies would have been greater.

Averting the risk of Poverty and Economic Insecurity.   Rural people behaved also in a rational way when they faced risky situations, and proved to be risk-averters: they migrated to town and to oil-rich Arab States to avoid poverty and unemployment; diversified crop cultivation to minimize the effects of output, income and market price fluctuations; and, in the absence of regular information about and insurance against weather uncertainty and market failures, they joined co-operatives as risk-pooling and a collective self-insurance arrangement. Likewise, and on their own initiatives, farmers have adjusted land tenancy arrangements in response to irrigation expansion and rising land profitability.   They replaced rent in kind by contractual fixed rent, especially in cotton-dominated agriculture (Chapter 6). Throughout North Africa, they also adopted sharecropping arrangements for the purpose of sharing the risk of crop-yields instability.   Many farmers also apply the Islamic principle of *takharog* (withdrawal of one inheritor in favor of others against compensation) to avoid subdivision of land as an arrangement for intergenerational property transfer (see Chapter 7, p.133).

Likewise, pastoralists and nomads -- surviving in an environment where weather risk is high -- have proved that they are also risk-averters. If they are not, the alternative is starvation.   They have customarily allocated their communal rights in land between herding, fallow, and growing food crops when rain permits.   They have formulated their own traditional code of conduct allowing for rotational grazing across inter-tribal pasture land according to variations in rainfall both in timing and in amount. In the absence of public investment in the provision of adequate water supply (as an insurance against drought), and owing to the lack of market insurance against death of livestock, their economic rationale in the allocation of resources proved to be effective for collective self-insurance. Nomads dig their own water holes *'hafir'* and wells, and accumulate livestock. They usually diversify their flocks into cows, camels and goats after they found from experience that herding requires labor specialization, the pattern of weather has changed, and that good years are becoming scarce. Without such rational and custom-determined arrangements, the disastrous effects of long droughts (and famine) would have been greater.

What Now? – Dilemmas

It seems that the era of concern about distributional and social welfare benefits in rural development has come to an end. In the 1990s, rural-agricultural development indicators will be significantly influenced by the extent of the shift from direct, extensive government intervention to market-determined changes, with a minimal role of the State in income transfer. Several political and developmental dilemmas emerge from this abrupt swing.

The General Dilemma

There is a general dilemma facing the policy-makers in the 1990s and beyond. Broadly speaking it is: Given the scarce means, how to satisfy rising expectations of rural people fostered by the governments' declared promises, and fattened by the politicians' rhetoric, and, at the same time, strive to realize the long-term national economic development objectives? Confronted with rising demands articulated vigorously by Islamic extremists and the writings of the intelligentsia, governments have tenaciously promised to quickly relieve the hardships and deprivation of the poor, and to give agriculture a top priority in resource allocation in order to bring about food security from domestic production. Can governments afford to meet these demands for immediate action to fulfill declared promises, and, at the same time continue their long-standing commitment to high spending on defence and industry? This dilemma is compounded by the fact that military and industrial establishments command power and have effectively competed with agriculture for scarce means at governments' disposal.

One can draw a parallel between these dilemmas and the dramatic shift in international development thinking and in the analytic reasoning behind policy choice. This shift was triggered around 1980 by developing countries' deep recession and their heavy foreign debts. This was, and still is, accompanied by attack and counter-attack between the advocates and opponents of an active role of governments in the management of the economy and in the redistribution of assets and income. The opponents have a powerful influence on the policy of international financing agencies, particularly IMF and the World Bank. They also receive strong ideological support from the rich Western donor countries. Their argued thesis is simple: no alternative to the free market as the basis of economic life. This forceful surge for anti-state redistributive intervention relies on a combination of financial strength and political pressure. Such a pressure has gained an increasing influence after the collapse of the Soviet Union which was a firm supporter of anti-poverty and pro-egalitarian rural development. Perhaps no other policy issue is more susceptible to these swift changes than the supply of cheap food to the poor, agrarian reform and redistributive equity. Likewise, no other socio-economic group is more vulnerable to the emerging policy prescriptions than the rural poor and their fellows in the cities.

Narrowed Policy Options

The 1950s and 1960s' wide range of policy options for the design of rural development strategy have gradually narrowed to a universally standardized package of macroeconomic reforms induced by the World Bank and IMF, the implementation of which is conditional upon the compliance of receiving-countries with the agreements.[4] Confronted with mounting economic difficulties in the early 1980s, North African governments, except Libya, decided to implement the prescribed adjustment-lending program, and agreed to comply with its conditions. This was not surprising. By 1986, Egypt, Morocco and Sudan were classified by the World Bank as highly indebted countries.[5] In fact, not all the causes of the experienced economic crisis originated in the mismanagement of national economies. A large part was rooted in the economic recession of the seven rich industrial countries. Its impact on the economies of North Africa was, and still is, through foreign debts, unprecedented rise in real interest rate, and a reverse trend in migrants' employment, exports and imports. Candidly, the real power behind the World Bank policy prescription lies in its control by this Group of Seven.[6] They possessed 49 percent of the voting power and 51 percent of the World Bank capital stock in 1986. This group is now the world stabilizer, setting world rules and providing developing countries with a wide range of things from high technology to sophisticated armament and wheat aid.

Given these realities of the international context, the policy-makers face dilemmas in the process of deciding on the path, scope, sequence and pace in introducing macroeconomic reforms. By accepting the conditions of the stabilization and structural adjustment package, governments are bound to be condemned by certain sections of their society for submitting to foreign pressures, labelled in some countries as 'imperialist agents' or 'new colonialism'. The political stability tends to be threatened by riots of the urban working class.[7] On the other hand, to exhibit their nationalistic attitude by rejecting the externally proposed set of measures, policy-makers may prolong the economic difficulties, losing their ability to repay foreign debts in the short-term, while the burden for their servicing increases and the balance of payments deteriorates further. The longer government action is postponed, the narrower their options become.

On implementing the adjustment package, the prices of consumer goods and services, together with exportable crops rise; the volume of agricultural exports and foreign-exchange earnings may increase and agricultural terms of trade improve, provided that the supply of required inputs does not decline and the world market demand for these crops does not deteriorate in quantity or price levels. Also, phasing out subsidies on fertilizers, irrigation-water rates, fuel and agricultural credit interest rates increases the costs of production and leads to higher consumer prices. Furthermore, this policy choice implies a redistribution of income away from food producers and food buyers (especially landless agricultural workers) to benefit producers of exportable cash crops; hence, with inflationary pressure, a downward trend in the purchasing power of a large section of the rural population for domestic goods and services.[8]

Another dilemma arises from deciding on major cuts in food subsidies and raising tax rates. Although these may immediately improve budget deficits, they, nevertheless, tend to worsen income distribution, and raise food prices. In the absence of an especially targeted food distribution program, these measures are likely to aggravate the already high incidence of malnutrition among the poor. Slashing public expenditure on rural roads and land reclamation, coupled with devaluation of domestic currency and raising tax rates increase unemployment, and reduce real incomes of many underpriviledged rural people. The rural groups who are bound to lose from macroeconomic reform programs are: those who are already poor and are at risk of becoming poorer. They comprize: wage-dependent landless workers who are net-food buyers; those losing jobs; and low-income rural population who would suffer from increased prices of goods and services. The quality of health and education services is likely to deteriorate as a consequence of cuts in current expenditure (school material, medicine, beds, vaccines and maintenance of buildings and facilities).

Empirical evidence suggests that policy choice is not uncomplicated. There are at least three problems. The first is the limitation of data on the likely effects of income transfer between wage earners and profit makers, on food consumption, real wages and incomes of the different categories of rural people. The second is the time-lag between changes arising from adjustment measures with respect to employment opportunities and price levels on the one hand, and the actual response of farmers as producers, consumers and traders, on the other. The third problem lies in the political domain. With IMF-World Bank's economic reforms, as with foreign aid, there is an important political dimension.[9]

To illustrate the political considerations, the Sudanese experience is relevant. In 1979, the government agreed with IMF, and later with the World Bank to implement a phased program with its tough conditions.[10] From 1979 to 1985, the government complied with the agreements' conditions, including a sharp devaluation of the Sudanese pound, a substantial reduction of wheat area in the Gezira irrigated land, and lifting government control of foreign-exchange. When President Nimeiri introduced, in 1983, the *Shari'a* Law (regarding Islamic principles on interest rates and exploitative practices in transactions), IMF suspended both the pledged loan and its financial support, on the grounds of the government's delay in the repayment of arrears. The United States government (USAID), which was always supporting the Nimeiri regime, paid the required amount on behalf of the Sudanese government. In April 1985, during Nimeiri's official visit to the U.S.A., he was ousted by a *coup d'état*. Subsequently, IMF informed the Sudanese government that Western donors were reluctant to provide financial support. With Sudan's increasing economic hardships after the 1984-1985 draught and famine, the, then, Minister of Finance (Sayed Abdelmageed) negotiated a new agreement which presented him with a dilemma. The agreement required no government control over foreign exchange, at a time when wealthy city merchants were active in accumulating foreign currency for its flight abroad. The Minister's rational response to his country's financial needs made him unpopular and cost him his job. Mounting political opposition to the IMF's callous position, forced the Minister to resign and, in February 1986, IMF terminated the agreement on the grounds that "Sudan was ineligible to use IMF resources" (Hussain: 1991, Appendix 5.1).

Long-Term Challenges and Dilemmas in Rural Development

We concentrated on the dilemmas related to the adopted economic adjustment reforms because, in North Africa as elsewhere, this subject preoccupies governments and people alike. It is within these reforms' framework that long-term policy issues in the development of rural-agricultural sector are to be settled. No task should command a higher priority for the policy-makers in the 1990s, than that of orientating the adjustment programs to reduce the degree of food insecurity and poverty incidence, or at least protect the poor from becoming poorer. The challenge is in devoting sufficient investment to agriculture for irrigation expansion and for increasing food production, and for raising the human quality of life in rural areas. This would, in turn, narrow the rural-urban income gap, and reduce the numbers of the poor. If the momentum is dissipated, the impact of the deep recession and that of the chronic imbalanced development would remain a problem of staggering dimensions.

The 1980's debt crisis and economic recession brought the structural weaknesses of North African countries into the open. But when programs and issues of structural adjustment preoccupied governments, little attention was paid to the effects on the long-neglected agriculture (in particular rain-fed subsector), and its poor landless cultivators. The same applies to the concern for malnutrition and land tenure problems. Policy-makers have been obsessed with the idea that reforming monetarist-fiscal policies and terms of trade would bring immediate accelerated growth, and the problems of the rural-agricultural sector, and those of the social welfare of rural people would be eventually resolved by the trickle-down path. Alas, this has not happened yet in most of the countries as Table 8.2 shows. We recognize that the initial degree of the recession gravity, as well as the mix and timing of implementation of economic reforms vary from country to country. However, the data provide an indication of the changes during the 1980s when economic reform policies were introduced, compared to the situation in 1960-1980.

There are different methods for an assessment of these policies. Clearly, a conclusive assessment is beyond the scope of this book. While we have explained the relevant aspects within the review of each country's experience (chapters 4, 5 and 6), the before-and-after approach is followed, for the purpose of a simplified inter-country comparison. We could not use Libya as a control country, because it has introduced, on its own account, some modifications mostly resulting from the Western countries' imposed trade sanctions, the sharp fall in world oil prices, and the heavy military costs incurred in the war with Chad.

The four macroeconomic indicators in Table 8.2 point to a worsening situation during the 1980s than the earlier period. Apart from a limited success in lowering inflation and preventing its rise, the growth rates of GNP per head and investment were disappointing. In all the five countries, the rates fell sharply in 1980-1990. The decline in *per capita* real income must have hit the poor harder than the non-poor. The rapid deceleration of investment implies a fall in savings and in the net-inflow of capital, both domestic and foreign. Likewise, with the exception of Morocco and Sudan, foreign-debt service as a percentage of value of exported goods and services has nearly doubled in 1990, compared to 1980.

A real dilemma faced the policy-makers, requiring a difficult choice. This was between reducing expenditure on basic needs (health, education and food subsidies) and the dire needs of the economies for foreign capital, included in the IMF-World Bank programs. They opted for the former as a deflationary measure, and the cuts in health and education were uneven. Sudan and Egypt cut the allocation to health services by 80 and 31 percent, respectively, while it was increased by 37 percent in Tunisia and by only 8 percent in Algeria. In varying degrees, the five countries protected education from cuts, and even increased its share in GNP. However, a close look at the distribution of public expenditure between rural and urban sectors and among the different levels of education manifests disparities, revealed by the Egyptian data.[11] Food subsidies were also cut from the high level of 10.7 percent of GDP in Egypt to 8.0 percent and in Morocco from the already low level of 3.8 percent to 2.3 percent. More recent data on Egypt show that this reduction raised the price of bread loaf *'baladi'* by 150 percent, rice by 147 percent and cooking oil by 193 percent in 1989-1990. This implies that the cost of purchased sources of calorie has, over 2 years, more than tripled, i.e. from 0.003 piasters to 0.01 piasters (per calorie-intake). Considering that nearly 60 percent of the calories come from bread and rice, and that the average inflation rate was 12 percent, this sharp rise has severely eroded the purchasing power of the poor (The World Bank: 1991, pp. 104-106).

TABLE 8.2   Comparison of Selected Indicators in Five Countries With IMF-World Bank Stabilization and Adjustment Programs Before and After 1980

Indicators	Sub-periods	Algeria	Egypt	Morocco	Sudan	Tunisia
1. *Macroeconomic Indicators*						
GNP per person, a.r.g.[a] %	1960-1980	3.2	3.4	2.5	-0.2	4.8
	1980-1989	0.0	2.8	1.3	-1.8	0.6
Gross Domestic Investment, a.r.g.[a] %	1970-1980	13.2	16.5	9.2	6.7	11.0
	1980-1990	-1.2	0.2	2.6	n.a	-3.1
Inflation, annual rate %	1970-1980	13.3	11.0	8.1	15.8	7.7
	1980-1990	6.6	12.0	7.2	n.a	7.4
Foreign debt service as % of exports earnings[b]	1980	27.1	14.8	32.7	25.5	14.8
	1990	59.4	25.7	23.4	5.8	25.8
2. *Food Security*						
Agric. GDP per working person, a.r.g [a] %	1970-1980	7.9	2.2	1.2	-0.1	3.5
	1980-1990	3.2	1.2	5.5	0.9	2.9
Food Production Per Person	1988-1990	96	118	128	71	87
Food imports dependency ratio [c] %	1969-1971	32.1	18.6	18.2	9.8	40.7
	1986-1988	70.7	45.2	28.1	14.5	59.3
3. *Government Expenditure*						
Health as % of GNP	1960	1.2	1.6	1.0	1.0	1.6
	1987	1.3	1.1	1.0	0.2	2.2
Education as % GNP	1960	5.6	4.1	3.1	1.9	3.3
	1989	9.4	6.8	7.3	4.8	6.3
Military exp. as % GDP	1960	2.1	5.5	2.0	n.a	2.2
	1989	1.9	4.5	4.3	6.0	4.9
Food subsidies as % of GDP	1982	n.a	10.7	3.8	n.a	2.8
	1985	n.a	8.0	2.3	n.a	n.a

Notes: a. arg stands for average annual rate of growth. n.a. not available.
b. The sum of repayments of the principal debt plus payments of interest made in foreign currencies and goods.
c. The percentage of food imports to total food available for domestic consumption.

(continues)

Table 8.2 (continued)
Sources: 1. *World Development Report* 1982 and 1992 (Indicators), except GNP per person 1980-1989 is from *Human development Report*, 1992, Table 24 of Indicators.
2. Agric. GDP per person in agriculture is from Table 3.1. Food production per head of total population is from *Production Yearbook*, FAO, Rome. Food imports dependency ratio is from *Human Development Report* 1992. Indicators, Table 13.
3. *Human Development Report*, 1991 and 1992, Indicators. Food subsidies in Egypt and Morocco are from Cornia *et al. Adjustment with a Human face*, Clarendon Press, Oxford, 1987, Table 3.9 and Tunisia from Radwan *et al*, 1991, Table 5.15.

But the real test in decreasing budget deficits was in military expenditure cuts. Policy-makers' response varied: slight cuts in Algeria and Egypt; and a sharp increase in Morocco and Tunisia. Even with these cuts, the outlay on imports of arms and the requirements of the military establishment absorbed, on average, about 7 percent of gross domestic product in 1989, compared to 3.6 percent in 1960, and to an average of 2.8 percent in all developing countries during the same period.[12] This obsession with the military meaning of security is manifested in the fact that the unproductive military expenditure per head of total population in 1989 remains enormous, in contrast to the principal human capital formation, health, as calculated below (multiplying GDP or GNP *per capita* by the corresponding percent share).

	Algeria	Egypt	Morocco	Sudan (1987)	Tunisia
health expend. per person US$	28.9	7.0	8.8	1.0	27.7
military expend. per person US$	31.0	27.9	39.3	28.9	54.6

These data underline an appalling disregard of human life and a distortion of the meaning of security. Security and socio-political stability cannot only be conceived in military and police force terms; it includes also economic security against economic shocks and unemployment. Of particular importance is food security at national and household levels stressed in Chapters 2 and 3. From the dawn of human history, command over food has been accounted as a major source of courage, power and security. In *Power - A New Social Analysis,* the philosopher Bertrand Russell says "The impulse of submission, which is just as real and just as common as the impulse to command, has its roots in fear" (1940, p. 18).

The three selected indicators of food security presented in Table 8.2 point to fear of the consequential effects of the high and increasing dependence of all the countries on foreign countries to feed their own people. They also suggest that economic reforms have been unable to alleviate food insecurity. Throughout this book, we have repeatedly signalled warnings of the consequences of falling food productivity and real income per person working in agriculture. We have also reiterated the harmful effects of continuing the heavy reliance -- for securing a substantial part of the countries' food requirements -- on imports against the use of scarce foreign exchange resources (except in the oil-based economies of Algeria and Libya). We have diagnosed this form of insecurity as being rooted in the interlinked problems of the fast, cancerous growth of population coupled with rapid urbanization on the one hand, and the continual disproportionate low allocation of government investment in both agriculture and rural people, on the other.

If national security is taken seriously as we suppose, country leaders and planners, in their Arab Summit meetings and national development plans,should give a high priority to regional co-operation in expanding irrigated land and increasing food production in order to build up not military weapons arsenals but food reserves. In this way, new vistas would open up for their rural-agricultural sectors. The postponement of such major decisions would bring about a continuous, grave imbalance of development, leaving the North African land, scarce water and rural labor underutilized, and food insecurity and absolute poverty unalleviated. These problems are not insoluble, if the dilemmas are faced squarely.

Changes for The Better

Let us conclude with a note of optimism. In 1992, there are indications of new developments towards the establishment of peace between long-time adversaries leading hopefully to cuts in defence expenditure, and to greater inflow of foreign aid and private capital which are likely to benefit agriculture. Importantly, the expected developments would bring about political stability, and transfer of large number of skilled youth from unproductive mandatory service in armed forces, to productive activities in the national economy.

We hope for the success of Israel and the Arab parties concerned in reaching an agreement acceptable to all present adversaries. A good possibility also exists for the settlement of local tribal feuds between Morocco and the Polisario, between Libya and Chad, and between the government of Sudan and the Southern rebels. Moreover, a source of fear requiring high military expenditure has been eliminated: communism has collapsed. The rural-agricultural sector is starving for the resources which are likely to be released as a result of the termination of these hostilities. There is a kindled hope that policy-makers put into action the inscription on the New York United Nations' building "Let swords become plows". In the next couple of decades, they have to be prepared for a different kind of war: the fight against mass illiteracy in rural areas and a battle of eradicating the fear from food insecurity and hunger to which governments have got coolly accustomed.

The 1980s was a decade of hardships and hope for quick improvements. But trade liberalization and other economic reforms, based on a revived faith in free-market mechanism, have not yet been able to solve the outstanding problems without a determination and political will to introduce corresponding reforms in the policy-making process. The painful economic reforms have an important claim to merit: plurality in economic activities within a free market has to be linked to a political plurality in policy formulation. From their past experience, planners, farmers and the middle-class intelligentsia have learnt that the larger the military establishment and centralized bureaucracy, the more powerful they have become, and the greater the risk of their choosing to dispense with genuine democracy of mass participation in policy formulation. And the greater the risk to run societies in their own interest. The rising anxiety for an anti-poverty rural development stewardship among non-governmental organizations, though being suppressed, is unstoppable. Policy-makers should at least listen to their views.

In a study like ours, the author, though born and bred in North Africa for nearly four decades, cannot prescribe a standardized rural development policy ready to implement. Rather, his professional capacity is to diagnose the problems, analyze policies, assess programs, raise broader questions related to the inquiry, and to warn against the evils he identifies. But how these are to be corrected, is the task of politicians, planners and truly elected representatives of the rural people. Judging by the bitter Algerian experience in 1990-1992, this process will be difficult. Genuine rural-agricultural development and true political participation, as always, demand difficult choices and prudent changes for the better. How can such change be brought about, is the challenge in the years ahead.

Notes

1. Imbalanced development refers to a departure of certain variables or elements, like investment rates, consumption, income, employment, institutional reforms, etc.) from a defined norm of balance. It may be viewed in terms of regions in a country (rural and urban or rainfed and irrigated areas within the rural sector) of the economy. Ideally a balanced type of development between two sectors of the economy (agriculture and industry) is where the identified elements (economic growth) or sectors are developed simultaneously and the real variables are growing at the same rate. Conventionally, the emphasis in development literature has been on inter-sectoral imbalance in efficiency of capital investment, employment, and in output growth differentials, irrespective of adverse distributional effects. Also, there is no agreement on what constitutes the imbalance and how to measure it. For a review of the concept and measurement, see chapter 2 "Concepts and Measurement of Uneven Development" in A.S. Balla *Uneven Development in the Third World: A Study of China and India,* Macmillan, London, 1992.

2. A land reform is complete if it meets the following conditions after implementation:

(i) The beneficiaries having direct access to landholding represent at least 50% of total agricultural households;

(ii) all or the majority of landless peasants are absorbed, thus leaving out none or a small fraction as landless workers;

(iii) the redistributed cultivable land amounts to over half the total; and

(iv) a low degree of inequalities in the distribution of land holdings is realized in terms of a Gini coefficient not exceeding 0.3; and

(v) a rising food production *per capita* of total population.

With regard to partial land reform policy, redistributive requirements are relative to the above listed four conditions. This implies a lower scale of the percentage of new landowners to total agricultural households and a correspondingly smaller proportion of cultivated land.

3. These remarks are based on the findings of a survey carried out in North African countries in 1982 - 1983 by the Near East and North Africa Association for Co-operatives and Agricultural Credit, published in Amman, Jordan, June 1983. The survey was conducted in each country by qualified nationals commissioned by the Association under the leadership of the late Ibrahim Abdullah, the then Secretary-General of this association

4. The reforms package includes stabilization and structural adjustment programs. The former refers to short-term policies, reducing budget and balance of payment deficits by way of raising the rate of interest, devaluating currency towards a more appropriate rate of exchange, and removing trade control, subsidies and quotas. Structural adjustment policies are of longer-term changes. They are designed to reform public-sector inefficiencies, correct the basic imbalances in the economy so as to improve the balance of trade and the efficient allocation of resources between the sectors of the economy.

5. In 1986 total foreign debts were as follows, in billions of US dollar, as listed in *World Development Report* 1988:

Algeria   17.9   Sudan 8.3
Egypt   28.6   Tunisia 6.0
Morocco 20.0 (1988)

6. The Group of Seven comprize Canada, U.S.A., U.K, France, Germany, Italy and Japan. The USA holds the greatest share of the World Bank's capital stock at 21 percent and 20 percent of its voting power, entitling her.to hold senior posts, including the Presidency.

7. From 1977 to 1989, severe riots rampaged major cities of Algeria, Egypt, Morocco, Sudan and Tunisia. They were provoked by austerity programs, particularly the sudden rise in food and fuel prices.

8. For judging the distributional impact of structural adjustment programs in developing countries, see (a) Cornia *et al, Adjustment with a Human Face,* 1987 and (b) Singh, 1987. For the methodology of analysis of effects on the rural poor, see Couvreur, 1989. See also 'The 1980s: Shocks, responses and the poor', in *World Development Report,* 1990.

9. See, for example, Hans Singer *'The Ethics of Aid',* Discussion Paper, Institute of Development Studies (IDS), Sussex, October 1984. An example of aid as a political tool, is the suspension of US aid to Egypt by the then Secretary of State, Henry Kasinger when President Sadat launched the October 1973 War against Israel. On signing the Camp David peace agreement in February 1979, Egypt became the largest American aid-receiver in the world after Israel, and all its balance of payment deficits was made up

by the USA and EEC generous aid. But Egypt was penalized by Arab donors whose aid, amounting to US $ 1.7 billion, was cut off after the Camp David agreements. Between 30 and 40 percent of that aid was for agriculture. Even worse, Sudan which refused to boycott Egypt after Camp David accord, was also penalized by Arab donors who were earlier commited to create in Sudan a 'bread-basket' and livestock ranches, the products of which were expected to be sufficient enough to meet most of the Arab States' demands. For a detailed account of this story of Arab aid, see Nonneman (1988, chapter 4).

10.   The conditions are: devaluation of the Sudanese pound; the creation of a free foreign-exchange market; severe cuts in government expenditures including subsidies; increasing tax and interest rates; elimination of export taxes; reforming the parastatals in the agricultural and industrial sectors; liberalization of production inputs, and the gradual repayment of foreign loans and debts.

11.   At 1980 constant prices, expenditure per student in primary and secondary education fell from 7.24 Egyptian pounds (LE) in 1980 to LE 2.4 in 1990, while that per University student slightly fell from LE 146.7 to LE 124. Moreover, the share of primary education which enrolled 40 percent of all was 20 percent of total expenditure on education, which contrasts sharply with University level, capturing 50 percent of total expenditure, while having only 10 percent of total students (*World Bank, Country Study, Egypt*: 1991, Annex J). Rural-urban disparity in health-expenditure cuts exists: endemic diseases hospitals in rural areas decreased by 6.3 percent and the number of beds by 21.2 percent in 1986-1990. At the same time, the number of hospitals *'Mostashfa A'amma'h'* in urban areas decreased by 5 percent, but the number of beds increased by 12.7 percent during the same period. At the aggregate level, mother and child care centers fell in numbers by 14.4 percent in 1986-1990 (*Statistical Yearbook*, 1991, Tables 5.5 and 5.6), in Arabic.

12.   On this point, a pioneer in research on development in the Arab Region says, "often the weapon arsenals and the armies are essentially meant for the internal security of the ruling juntas or regimes, or as a warning to neighbours, often referred to as 'sister countries', not in reality for a genuine external enemy" (Yusif A. Sayigh: 1991, p. 23). The high military expenditure in North Africa, is in contrast to the world preserver of stability, the western rich industrialised countries, whose expenditure is only around 4% of GDP.

Appendixes

Appendix 1

Cereal Production Instability in Sudan

TABLE A.1 Annual Deviation in Total Cereals Yields, Sudan, 1960 - 1990

Year	Yield Ton/Hectare	Deviation from Mean	Year	Yield Ton/Hectare	Deviation from Mean
1960	0.76	+0.113	1976	0.62	- 0.027
1961	0.91	+0.263	1977	0.67	+0.023
1962	0.81	+0.163	1978	0.74	+0.093
1963	0.88	+0.233	1979	0.54	- 0.107
1964	0.78	+0.133	1980	0.66	+0.013
1965	0.69	+0.043	1981	0.79	+0.143
1966	0.60	- 0.047	1982	0.49	- 0.157
1967	0.91	+0.263	1983	0.45	- 0.197
1968	0.67	+0.023	1984	0.29	- 0.357
1969	0.80	+0.153	1985	0.60	- 0.047
1970	0.74	+0.093	1986	0.57	- 0.077
1971	0.78	+0.133	1987	0.35	- 0.297
1972	0.62	- 0.027	1988	0.61	- 0.037
1973	0.59	- 0.057	1989	0.35	- 0.297
1974	0.61	- 0.037	1990	0.47	- 0.177
1975	0.71	+0.063			

No. of years = 31.
Mean value of yield = 0.647 Ton/ha.
No. of years in which the yield is below the mean value = 15.

Notes: Total cereals are sorghum, millet, maize and wheat.

Sources: From 1970 to 1985, yields are taken from *Processed Statistics Series* Vol. 1, FAO, Rome. From 1986 to 1990 are from *Production Yearbook*, Vol. 42, 43 and 44, FAO, Rome.

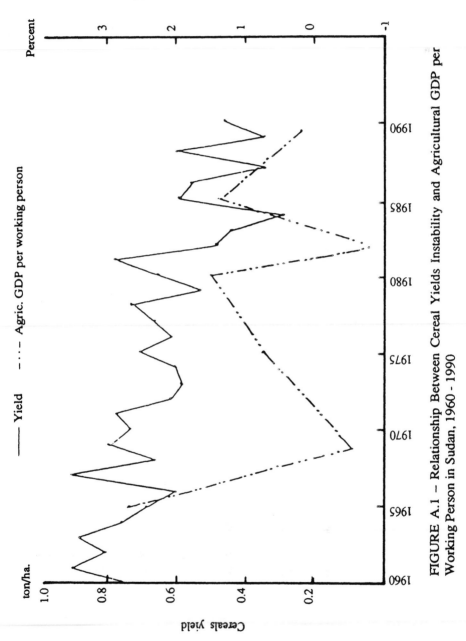

Agric. GDP annual growth rates per working person

Percent

—— Yield   – · – · – Agric. GDP per working person

FIGURE A.1 – Relationship Between Cereal Yields Instability and Agricultural GDP per Working Person in Sudan, 1960 - 1990

Appendix 2

Fragmented holdings in Newly Reclaimed Land, Egypt

Appendix TABLE 2 Number of Fragmented Plots and Distance Between Plots of the
Same Holding in the Land Settlement Scheme in Newly Reclaimed
Land, Egypt, 1974

Farm Area	Total No. of Plots	Number of plots by Distance between one plot and another, meters			
		Less than 500	500 to less than 1,000	1,000 to less than 2,000	Over 2,000
El-Manshia	456	81	109	211	55
El-Sabeen	305	81	98	121	5
M. Yazid	245	107	90	40	8
M. Kamel	274	33	44	150	47
El-Nasr	408	72	65	145	126
El-Sa'ada	699	263	404	32	3
Edko/5	396	66	170	132	28
Hares/2	504	3	90	298	113
El-Kawmeya	891	490	290	111	-
El-Horreyn	292	132	100	60	-
Total	4,473	1,328	1,460	1,300	385

Notes: This planned scheme is situated in North-Western Nubaria about 50 kilometers south of the Mediterranean coast. The study reveals that the plot location was suitable in only four farming areas out of the ten studied. In 83% of the distributed land in the other six farming areas the location appeared to be haphazard and badly located which led to great disadvantages, and brought about a situation detrimental to the aims of the "multi-plot" holding system. The distance between the settler's house and his holding, which should not exceed 2 kilometer's, was found in only 61% of total number of settlers. It was also found that 18% of the surveyed settlers lived in houses more than 5 kilometers far from their holdings.

Source: Salah Wazzan, "Evaluation of Agricultural Land Use Consolidation of Individual Holdings in Newly Reclaimed Lands". United Nations UNDP project 71, Alexandria, May 1974, in Arabic.

Appendix TABLE 3   Expansion of Collective Co-operatives in Algeria, 1973-1977

Period	CAPRA[a] No.	%	CAEC[b] No.	%	GPMV[c] No.	%	Others No.	%	Total Number of Co-ops. 100%
By July 1973	1,391	53	825	32	398	15	--	--	2,614
December 1974	1,748	60	601	20	572	20	86	4	2,921
Total by the end of 1977	4,203	72	528	9	930	16	180	3	5,841

Notes: Percentages are rounded, adding to 100 in a row.
a. CAPRA = Coopèratives agricoles de production de la révolution agraire.
b. CAEC = Cooperatives agricoles d'exploitation en commun.
c. GPMV = Groupement pre-coopèrative de mise en valeur. They represent the first stage in collective farming.

Source: Ministry of Agriculture and Agrarian Reform: *Enquête sur les coopèratives*, pp. 13-14 and *Annuaire Statistique de l'Algèria*, 1977.

Appendix 4

Nutritional Status in North Africa, 1969-1985

TABLE A.4.1 Calories Per Person Per Day, 1969-1985

	1969-1971	*1974-1976*	*1980-1982*	*1984-1986*
Algeria	1,825	2,168	2,682	2,687
Egypt	2,499	2,692	3,122	3,313
Libya	2,367	3,468	3,670	3,617
Morocco	2,424	2,573	2,727	2,738
Sudan	2,115	2,102	2,312	2,941
Tunisia	2,271	2,610	2,774	2,942

Source: *The State of Food and Agriculture*, 1989, Annex Table 16, FAO, Rome.

TABLE A.4.2 Rural-Urban Incidence of Undernutrition in Egypt, Morocco and Tunisia, 1975-1988

Year Indicator and Country		*Rural*	*Urban*	*National*
Egypt				
1982[a]	average calories per day/person	2,980	2,742	2,843
1984[b]	persons below 2,000 calories per day %	38.5	33.1	35.3
1988[b]	Children under 5 Acute			7.0
	Chronic/moderate			24.1
	Anemaemia			50.7
Morocco				
1985[a]	average calories per day/person	2,607	2,266	2,460
Tunisia				
1985[a]	average calories/day/person	2,424	2,158	2,276
1975[a]	Children under 5 Acute			2.0
	moderate			25.1
	Aneamia	28.4	33.3	29.9

Note: Acute undernutrition equals weight for height less than 85% of standard. Moderate or chronic equals height for age less than 90% of standard.

Sources: a. Nutrition Country Profile, Egypt (1987) Morocco (1989), Tunisia (1989), FAO, Nutrition Division, 1987. b. World Bank Country Study, Egypt, 1991, Tables 2.5, 2.6 and 2.8.

Appendix 5

Studies Prepared By the Author on North Africa and
The Near East*, 1965-1990

1965. "The role of land policy in agricultural production and income distribution in the Near East". A paper presented at the Regional Seminar on Land Policy, Tripoli, Libya, October 1965. Published in El-Ghonemy, ed. 1967. *Land Policy in the Near East*. Rome: FAO.

1966. "Land Reform and Economic Development in the Near East". A paper presented at the World Land Reform Conference, Rome, Italy, July 1966, and published in *Land Economics*, Vol. XLIV, No. 1, pp. 36-49, February 1968.

1970. "RegionaL Analysis of Agrarian Reform in the Near East in the Context of the First and Second Development Decades". A study prepared for FAO Special Committee on Agrarian Reform.

1971. "Land Settlement as a Developmental Instrument in the Near East". A paper presented at the Regional Workshop on Human Settlement in New Lands organized by FAO and Ford Foundation, Cairo, September 1971.

1972. *Population and Food Problems Related to Rural Development in the Near East Region*. A study presented at the Regional Seminar organized by FAO and United Nations Fund for Population Activities (UNFPA), Cairo, Egypt, December 1972, published by UNFPA and The Institute of Statistical Studies, Cairo University, 1976.

1974. "Integrated Rural Development in the Near East: Conception and empirical evidence". A study presented at the 12th FAO Regional Conference of Ministers of Agriculture of the countries in the Region, Amman, Jordan, September, 1974.

1976. "Agrarian Reform and Farm-size Productivity in the Near East". A paper presented at the 13th FAO Regional Conference of the countries' Ministers of Agriculture, Tunis, Tunisia, October 1976.

1979. "Agrarian Reform and Rural Development in the Near East. Analysis of policies since the mid-1960s and the task ahead in the 1980s". A study prepared for the World Conference on Agrarian Reform and Rural Development, Rome, July 1979.

1984. *Economic Growth, Income Distribution and Rural Poverty in the Near East*. FAO, Rome, 1984, in Arabic and English.

1989. "Land Tenure and Policy in North Africa". A study prepared for the United Nations Economic Commission for Africa, Addis Ababa.

1990. "Land Tenure Systems and Rural Poverty in the Near East and North Africa". A study prepared for the International Fund for Agricultural Development (IFAD), Rome.

---

* The Near East comprizes all North African countries, the rest of Arab States, and Cyprus, Turkey, Iran, Afghanistan and Pakistan.

Appendix 6

Regional Institutions For Co-operation
Between Arab States

Kuwait Fund for Arab Economic Development (1961)

Arab Fund for Economic and Social development (AFESD, 1971)

Arab Common Market (1971)

Abu Dhabi Fund for Economic Development (1971)

Islamic Development Bank (1974)

Arab Investment Corporation (1974)

Arab Authority for Agricultural Development (1976)

Saudi Fund for Development (1974)

Iraqi Fund for Technical Assistance among Arab Countries (1974)

Arab Authority for Agricultural Development (1976)

Islamic Development Bank (1975)

Near East and North Africa Association for Agricultural Credit and

Co-operatives (1978)

Arab Centre for the Study of Arid and Dry Lands

Near East Regional Commission for Economic and Social Policy (1987)

---

* Year of establishment is between brackets.
These institutions are chosen for the relevance of their work to agricultural/rural
development.

Glossary *

amlāk is the plural of *melk:* full ownership. *melkiyya:* property.

amn ghiza'i or ghidhā'i: food security.

awqāf is the plural of *waqf:* property not subject to normal transactions in private property, but its revenue is assigned to either family members (*waqf ahli*) or to a religious and social purposes (*waqf khairi*). In the Maghreb countries the term *habous* (singular *habs*) is widely used.

felāha: agriculture in the Maghreb, and *zira'a* in Libya, Egypt and Sudan.

fellaheen: an Arabic plural name of *fellah*, traditionally given to the social classes of peasants and hired workers in settled agriculture. In pasture lands of herdsmen and nomads (bedouin), they are termed *bad'w ro'ah*. *badawiy* is singular and also an adjective.

qadi: judge.

qardh hasan: interest-free loan.

qariah: village; in some areas it is named *dow'war*.

reef: countryside or rural areas. *reefi* (adjective), hence *tanmiyya reefiyyah:* rural development

shaikh (or Shaykh): bedouin chief, moslem scholar (also faqih), and village headman. Its plural is *mashayikh* or *shiyoukh*.

ulamā is the plural of aalem: scholarly man, an authority on Islamic matters.

zakāt: one of the five pillars of Islamic belief, meaning almsgiving to the poor (*al-foqarā*) and needy (*wal-mohtageen*).

---

* The sole purpose of this transliteration of basic Arabic terms, used in the book, is to enable the reader to understand their usage in North Africa as well as in relevant publications. The Egyptian pronounciation of 'g' replaces 'j' as in *jomlah* and *ijmali*, meaning total, and *Jomhouriyya:* Republic. Also names of persons, though spelled in the same way throughout North Africa, are spelled differently in English forms (e.g. Mohammad, Mohammed, Muhammad, Mehmed). In this book, we have represented the Arabic words in their English forms, as they are pronounced in each country, and the names of authors, as they spell them.

Acronyms and Abbreviations

ABS   Agricultural Credit Bank of Sudan (Khartoum)

CAPMAS   Central Agency for Public Mobilization and Statistics (Cairo)
CIMMYT   International Maize and Wheat Improvement Center (Mexico City)
CPI   Consumer Price Index

ECA   Economic Commission for Africa (United Nations Addis Ababa)
EEC   European Economic Community (Brussels)

FAO   Food and Agriculture Organization of the United Nations (Rome)

GDP   Gross Domestic Product
GFCF   Gross Fixed Capital Formation
GNP   Gross National Product

HES   Household Expenditure Survey
HYV   High Yielding Varieties of Seeds

ICARDA   The International Center for Agricultural Research in Dry Areas (Aleppo, Syria)
IDS   Institute of Development Studies at the University of Sussex (Brighton, U.K)
IFPRI   International Food Policy Research Institute (Washington, D.C.)
ILO   International Labor Organization of the United Nations (Geneva)
IMF   International Monetary Fund (Washington, D.C)
INS   Institut National des Statistiques (Tunis)

LDC   Less Developed Countries or Developing Countries
LE   Egyptian Pound
LS   Sudanese Pound

UNCTAD   United Nations Conference on Trade and Development (Geneva)
UNDP   United Nations Development Program (New York)
UNESCO   United Nations Education, Scientific and Cultural Organization (Paris)
UNFPA   United Nations Fund for Population Activities (New York)
UNICEF   United Nations International Children's Emergency Fund (New York)
USAID   United States Agency for International Development (Washington, D.C)

WFP   World Food Program of the United Nations (Rome)
WHO   World Health Organization of the United Nations (Geneva)

Bibliography

Abbott, Philip. 1991. "The Economics of Wheat Production in Morocco", *Journal of Agricultural Economics* 42: 23-32.

Abd al-Kader, Ali. 1959. "Land Property and Land Tenure in Islam" *The Islamic Quarterly* 5:4-11.

Abdalla, Abbas. 1986. "Water Supply Factor in Sudan", in Antoine Zahlan and Wadie Magar, eds. *The Agricultural Sector of Sudan*. London: Ithaca Press.

Abd ar-Rahim. 1990. "Sudan History" in *The Middle East and North Africa*. London: Europa Publications.

Abdel-Khalek, Mohsen. 1971. "Agrarian Reform in Egypt, a Field Study in Two Areas 1953-63". Unpublished Ph.D. Thesis, University of London.

Abu-Sheikha, Ahmad. 1983. "Towards the Alleviation of Rural Poverty in the Sudan". Poverty studies series: No. 1, mimeographed. Rome: FAO.

Abu-Zahra, Mohamad. 1963. *Ahkām al-tar'ikaat wal Mawāreeth* (Laws of Inheritance and Legacies). Cairo: Dar al-Fikr al-Arabi.

Adam, Farah. 1981. Analysis of Existing Landowner-Tenant Relationships. Khartoum: University of Khartoum.

_____. 1987. "Agrarian Relations in Sudanese Agriculture, in Elfatih Shaaeldin, ed., The Hague: Institute of Social Studies.

Adams, Richard. 1985. "Development and Structural Change in Rural Egypt, 1952 - 1982", *World Development* 13: pp 705-23.

Ahmad, Khurshid. ed. 1981. *Studies in Islamic Economics*, papers presented at the International Conference on Islamic Economics held in Makka, February 21-26, 1976. Glasgow, Scotland: Robert MacLehose.

Ahmad, Saad. 1986. "Rainfed Mechanized Farming in Southern Gedaref", in Zahlan and Magar, eds. *The Agricultural Sector of Sudan*, London: Ithaca Press.

Albeltagy, Adel. 1987. "Land Reclamation Schemes and Expanding Employment Opportunities", in *Proceedings of Symposium on Development of Employment Opportunities*, Cairo, 20-22 December, in Arabic and English.

Alderman, Harold and Joachim von Braun. 1984. *The Effects of The Egyptian Food Ration and Subsidy System on Income Distribution and Consumption*. Washington, D.C: IFPRI.

Alexandratos, Nikos. ed. 1988. *World Agriculture Towards 2000. An FAO Study*. London: Pinter, Belhaven Press.

Algeria, Ministry of Planning. 1985. *Deuxieme Plan Quinquennal, 1985-1989, Rapport Général*.

_____. *Tableaux de l'Économie Algèrienne*, several issues.

_____, 1975. Commission Nationale de la Révolution Agraire. *Révolution Agraire: Textes Fondamentaux*.

Ali, Ahmad. 1986. "Finance and Credit", in Zahlan and Magar, eds. *The Agricultural Sector of Sudan*. London: Ithaca Press.

Ali, Taisier. 1988. "The State and Agricultural Policy", in Tony Barnett and Abbas Abdel-karim, eds. *Sudan: State Capital and Transformation*. London: Croom Helm.

Allan, J.A., Keith McLachlan and Edith Penrose, eds. 1973. *Libya: Agriculture and Economic Development*. London: Frank Cass.

Al-Quddafi, Muammar. 1976. *al-Kitab al-Akhdar*. (The Green Book). Tripoli: Government of Libya, Ministry of Information, in Arabic and English.

Artin, Y. 1885. *La Propriété Fonciére En Egypt* (1883), translated and published in London as *The Right of Landed Property in Egypt*.

Awad, Mohamad. 1987. "The Evolution of Land Ownership in the Sudan", in Elfatih Shaaeldin, ed. *The Evolution of Agrarian Relations in the Sudan*. The Hague: Institute of Social Studies.

Baer, Gabriel. 1962. *A History of Landowners in Modern Egypt 1800-1950*. Oxford: Oxford University Press.

Balla, A. 1992. *Uneven Development in The Third World: A Study of China and India*. London: Macmillan Press.

Bardhan, Pranab. 1984. *Land, Labor and Rural Poverty*. Delhi: Oxford University Press.

_____. ed. 1989. *The Economic Theory of Agrarian Institutions*. Oxford: Clarendon Press.

Barnett, Tony and Abbas Abdel-Karim, eds. 1988. *Sudan: State, Capital and Transformation*. London: Croom Helm.

Bedrani, Slimani. 1987. *Les Pasteurs et Agro-pasteurs au Maghreb*. Rome: FAO.

Bennamane, Aissa. 1983. "Agrarian Transformation in Algeria". A study prepared for FAO, Rome, mimeographed.

Bigsten, Arne. 1983. *Income Distribution and Development: Theory, Evidence and Policy*. London: Heinemann.

Binswanger, Hans, John McIntire and Chris Udry. 1989. "Production Relations in Semi-ardi African Agriculture", in Pranab Bardhan, ed. *The Economic Theory of Agrarian Institutions*. Oxford: Clarendon Press.

_____. and Mark Rosenzweig. 1986. "Behavioral and Material Determinants of Production Relations in Agriculture". *The Journal of Development Studies*, 22: 503-29.

Biswas, Asit. 1991. *Land and Water Management for Sustainable Agricultural Development in Egypt: Opportunities and Constraints*. A study prepared for FAO, project TCP/EGY/0052. Rome: FAO and Egypt's Ministry of Agriculture.

Booth, Anne and P. Sandrum. 1985. *Labor Absorption in Agriculture*. New York: Oxford University Press.

Bouaita, Ahmed and Claudine Chaulet. 1990. "Mechanization and Agricultural Employment in Arid and Semi-arid Zones of Morocco: The case of Upper chaouia", in Dennis Tolly, ed. *Labor, Employment and Agricultural Development in West Asia and North Africa*. Dordrecht, the Netherlands: Kluwer Academic Publishers.

Bouderbbala, N. and Paul Pascon. 1966. *La Question Agraire au Maroc*. Rabat: Bulletin Economique et social.

Burks, J. and C. Sinclair. 1980. *Arab Manpower: The Crisis of Development*. London: Croom Helm.

_____. 1980. *International Migration and Development in the Arab Region*. Geneva: ILO.

Cater, Nick. 1986. *Sudan: The Roots of Famine*. Oxford: Oxfam.

Chebil, Mohsen. 1967. "Evolution of Land Tenure in Tunisia in Relation to Agricultural Development Programs" in Riad El-Ghonemy, ed. *Land Policy in the Near East*. Rome: FAO.

Chenery, Hollis, Montek Ahluwalia, Clive Bell, John Duloy, and Richard Jolly. 1974. *Redistribution with Growth*. Oxford: Oxford University Press.

CIMMYT. 1990-1991 *World Wheat Facts and Trends*, Mexico City.

Cleaver, Kevin. 1982. *The Agricultural Development Experiences of Algeria, Morocco and Tunisia*, World Bank Staff Working Papers, No. 552. Washington D.C.: The World Bank.

Cornia, Giovanni. 1985. "Farm size, land yields and the agricultural production function: An analysis for fifteen developing countries". *World Development*, 13, 4: 513-34.

_____, Richard Jolly, and Frances Stewart, eds. 1987. *Adjustment With a Human Face: Protecting the Vulnerable and Promoting Growth*. A UNICEF study, Vol. 1. Oxford: Clarendon Press.

_____, *et al.* 1988. *Adjustment with a Human Face*. Vol. II, Country Case Studies. A UNICEF Study. Oxford: Clarendon Press.

Cox, Terry. 1986. *Peasants, Class and Capitalism*. Oxford: Clarendon Press.

Couvreur, Yven. 1989. "The Impact of Adjustment Programs on Poverty and Income Distribution in Agriculture", unpublished M.Sc Thesis, University of Oxford.

Commander, Simon. 1987. *The State and Agricultural Development in Egypt since 1973*. London: Ithaca Press.

Daden, Mohamad. 1978. *La Subventions dans le cadre du Code des Investissements Agricoles et la Conception Dualiste de la Modérnisation car de Gharb*. Rabat: Institut Hassan II.

de Janvry, Alain and Elisabeth Sadoulet. 1989. "Agrarian Structure, Technological Innovations, and the State", in Pranab Bardhan, ed. *The Economic Theory of Agrarian Institutions*. Oxford: Clarendon Press.

de Waal, Alexander. 1989. *Famine That Kills, Darfur, Sudan. 1984-1985*. Oxford: Clarendon Press.

_____. 1991. "Emergency Food Security in Western Sudan: What is it for?", in Simon Maxwell, ed. *To Cure All Hunger: Food Policy and Food Security in Sudan*. London: Intermediate Technology Publications.

Diab, Mohamad and David Evans. 1991. "Food Security, Income Distribution, and Growth in Sudan", in Simon Maxwell, ed.

Economist International Unit, (The) 1991. *Libya, Country Profile*. London: Business International, The Economist.

Egypt, Ministry of Agriculture. *al-Iqtissād al-Zira' ie* (Bulletin of the Agricultural Economy, several issues, Dokki, Cairo.

_____. CAPMAS, *Statistical Yearbook*, several years. Cairo, in Arabic and English.

_____, Ministry of Finance. 1946. *Taqdeer Darā' eb al-Arādy al Zirai' eya* (Land tax Procedures) Law No. 53 1939 and its regulations; and 1949. *Maslahat al-Amlāk al-Amereya Inshā' oha wa A' amāloha*, (both documents are in Arabic).

_____. 1988. Ministry of Agriculture, *The Results of the Agricultural Census of 1981-1982*, in Arabic.

El-Firgani, Nader. 1984. *al-Higra Ela al-Naft* (The migration to oil). Beirut: Center for Arab Unity Studies.

El-Ghonemy, M. Riad. 1953. "Resource Use and Income in Egyptian Agriculture Before and After Land Reform", unpublished Ph.D Thesis, North Carolina State University, Raleigh, N.C, U.S.A.

_____. 1965. "The Development of Tribal Lands and Settlements in Libya", *Land Reform*, An FAO Journal. Rome: FAO.

_____. ed. 1967. *Land Policy in the Near East*. Rome: FAO.

_____. 1968. "Land Reform and Economic Development in the Near East", *Land Economics* XLIV (1): pp. 36-49.

_____. 1968. "Economic and Institutional Organization of Egyptian Agriculture Since 1952", in P.J. Vatikiotis, ed. *Egypt since the Revolution.* London: George Allen and Unwin, Ltd.

_____. 1979. *Agrarian Reform and Rural Development in North Africa and the Near East.* Cairo: FAO Regional Office for the Near East, in Arabic and English.

_____. 1990. *The Political Economy of Rural Poverty: The case for land reform.* London and New York: Routledge.

_____. 1992. "The Egyptian State and Agricultural Land Market: 1810-1986". *Journal of Agricultural Economics* 43(2): pp. 175-90.

_____, Godfrey Tyler and Khan Azam. 1986. *Yemen Arab Republic: Rural Development Strategy and Implementation.* Report of the United Nations ESCWA Mission. Baghdad: ESCWA/FAO Agriculture Division.

El-Gritly, Aly. 1977. *Twenty Five Years: An Analytical Study of Economic Policy in Egypt, 1952-1977.* Cairo: al-Hai'ah al-Ammah lil-Kitab (in Arabic).

El-Imam, Mahmoud. 1962. *A Production Function for Egyptian Agriculture,* 1913-1955, Study No. 259, mimeographed. Cairo: Institute of National Planning.

El-Issawy, Ibrahim. 1982. "Interconnections between Income Distribution and Economic Growth in the Context of Egypt's Economic Development", in Gouda Abdul-Khalek and Robert Tignor, eds. *The Political Economy of Income Distributio in Egypt.* New York: Holmes and Meier Publishers.

El-Jawhary, Hamid. 1967. "Land Settlement Policy and Projects in Libya", in Riad El-Ghonemy, ed., *Land Policy in the Near East.* Rome: FAO.

El-Wifaty, Beshir. 1978. "Evaluation of Land Settlement Programs in Libya". A study prepared for FAO Regional Office for the Near East, mimeographed.

Faisal Islamic Bank. 1979. *Faisal Islamic Bank: Its Objectives and Operational Methods.* Khartoum: Faisal Islamic Bank.

_____. 1983. *Report of the Board of Directors.* Khartoum: Faisal Islamic Bank (in Arabic).

FAO. 1983. *Review of Food Consumption Survey 1981.*

_____. 1985 and 1990. *Country Tables: Basic Data on the Agricultural Sector.*

_____. *Production Yearbook.* Several years.

_____. 1985. *The Fifth World Food Survey.*

_____. 1986. *Atlas of African Agriculture.*

_____. 1987. *Food Aid in Figures.*

_____. 1987. "The Effects of Land Tenure and Fragmentation of Farm Holdings in Agricultural Development". A study for the Committee on Agriculture. Mimeographed.

_____. 1981, 1989, 1991. *The State of Food and Agriculture.*

Gadalla, Saad. 1962. *Land Reform in Relation to Social Development in Egypt.* Jefferson city: University of Missouri Press.

Gaitskell, Arthur. 1959. *Gezira: A Story of Development in the Sudan.* London: Faber and Faber.

Ghai, Dharam, and Samir Radwan, eds. 1983. *Agrarian Reform and Rural Poverty in Africa,* Geneva: International Labor Office.

Gibb, H. and Harold Bowen. 1950. *Islamic Society and The West.* London: Oxford University Press.

Griffin, Keith. 1976. "Income Inequality and Land Redistribution in Morocco", in *Land Concentration and Rural Poverty,* London: Macmillan Press.

Hansen, Bent. 1966. "Marginal Productivity, Wage Theory and Subsistence Wage Series in Egyptian Agriculture", *Journal of Development Studies* 2: pp. 367-407.
_____ and Girgis Marzouk. 1965. *Development and Economic Policy in the U.A.R. (Egypt).* Amsterdam: North-Holland Publishing Co.
Hodnebo, Kjell. 1981. *Cotton, Cattle and Crises : Productikon in East Equatoria Province, Sudan*, 1920-1950. Fantoft, Norway: The C.H.R. Michelsen Institute.
Hicks, Norman. 1980. *"Economic growth and human resources"*, World Bank Staff Working Papers, No. 408, Washington.
_____, and Paul Streeten. 1981. "The search for a basic needs yardstick" in Paul Streeten, *Development Perspectives*. London: Macmillan.
Hussain, Mohammed. 1991. "Food Security and Adjustment Programs: The conflict", in Simon Maxwell, ed. London: Intermediate Technology Publication.

Ibrahim, Ahmad. 1982. 'Impact of Agricultural Policies on Income Distribution', in Gouda Abdel-Khalek and Robert Tignor, eds., *The Political Economy of Income Distribution in Egypt*. New York: Holmes and Meier Publishers.
IDS. 1988. *Sudan Food Security Study: Final Report*. Brighton: IDS, University of Sussex.
IFAD. 1988. "Report of the Special Programming Mission to Sudan", mimeographed. Rome: IFAD.
ILO. 1984. *Labor Market in the Sudan*. Report of the ILO mission to Sudan. Geneva: ILO.
Instituto Agricola Coloniale. 1947. *La Colonizatione Agricola della Tripolitania e della Cirenaica*. Florence, Italy.
Issawi, Charles. 1982. *An Economic History of the Middle East and North Africa*, New York: Columbia University Press.

Johansen, Baber. 1988. *The Islamic Law on Land Tax and Rent*. London: Croom Helm.
Johnston, Bruce and William Clark. 1982. *Redesigning Rural Development: A Strategy Perspective*. Baltimore: The Johns Hopkins University Press.
IMF. *Government Finance Statistucs Yearbook*, several years.
_____. 1980. *The Experience of Sudan*. Washington, D.C: IMF.
_____. 1986. *Current Economic Situation in Sudan*. Washington, D.C: IMF.

Keen, David. 1991. "Targeting Emergency Food Aid: The Case of Darfur, Sudan in 1985", in Simon Maxwell ed., *To Cure All Hunger*.
Khedr, Hassan. 1989. "Public Expenditure and Agricultural Taxation: The Case of Egypt". A study prepared for FAO, Rome (mimeographed).
Khrouz, Driss and Moha Marghi. 1990. "Implications for Technological Change for Labor and Farming in the Karia Ba Mohamad District, Morocco", in Dennis Tolly, ed., *Labor, Employment and Agricultural Development in West Asia and North Africa*. Dordrecht, the Netherlands: Kluwer Academic Publishers.

Lazarev, G. 1977. "Aspects du Capitalisme Agraire au Maroc Avant Le Protectorat", in Bruno Etienne, ed., *Les Problémes Agraires au Maghreb*. Paris: Centre National de la Recherche Scientifique.
Lele, Uma. 1975. *The Design of Rural Development: Lessons From Africa*. Baltimore: The Johns Hopkins University Press.
Libya, Secretariat of Agriculture. 1978. *Drassa 'an Siyasāt wa Barāmeg al-Islāh al-Zira'i wal tanmeya al-rifeyya* (a study on Agrarian Reform and Rural Development in Libya), Tripoli (mimeographed).

_____, Secretariat of Planning: *Khetat al-tahawol al iqtisādi wal-igtima'ee wa taqueemiha* (1976-1980, Plan for Socio-economic Transformation). 1976-1980, Triploi; and *Summary of the 1980-1985 Development Plan.*

Little, I.D. 1965. *A Critique of Welfare Economics.* Oxford: Oxford University Press.

Livingstone, Ian. 1984. *Pastoralism.* Rome: FAO.

Magar, Wadie. 1986. "The White Nile Pump Schemes", in Antoine Zahlan, ed., *The Agricultural Sector of Sudan.* London: Ithaca Press.

Maxwell, Simon, ed. 1991. *To Cure All Hunger: Food Policy and Food Security in Sudan.* London: Intermediate Technology Publications.

Mead, Roland. 1967. *Growth and Structural Change in the Egyptian Agriculture.* Homewood, Illinois: Irwin, Inc.

Morocco, Ministry of Agriculture. 1973/1974. *Recensement Agricole.*

_____. 1987 *L'Agriculture en Chiffres.*

_____. 1989 *al- felaha fi tanmiyya mostamerah (Agriculture in continual development)*

_____. 1990 Rapport sur l'avancement du Développement rural.

_____. 1960-1964 Ministry of Planning, First Development Plan.

_____. 1984-1985 *Consommation et Dépenses des Ménages Prémiers Résultats,* Vol. 1: Rapport de synthèse

Nonneman, Gerd. 1988. *Development Administration and Aid in the Middle East.* London: Routledge.

Nouvele, J. 1949. "La Crise Agricole de 1943-1946 au Maroc et ses Conséquences Économiques et Socialies", *Revue de Géographie Humaine* I: pp. 87-89.

Nozick, Robert. 1974. *Anarchy, State, and Utopia.* Oxford: Basil Blackwell.

O'Brien, Patrick. 1969. "The Long Term Growth of Agricultural Production in Egypt: 1821-1962" in P.M. Holt, ed., *Political and Social Change in Modern Egypt.* London: Oxford University Press.

Osman, Omar. 1987. "Some Economic Aspects of Private Pump- Schemes", in Elfatih Shaaeldin, ed. *The Evolution of Agrarian Relations in the Sudan.* The Hague: Institute of Social Studies.

Owen, Roger. 1969. *Cotton and the Egyptian Economy, 1820- 1914.* Oxford: Clarendon Press.

Oxfam and UNICEF. 1986. "A Report on the Nutritional Status of 3,183 Children in Kordofan Region". Oxfam Field Office, Khartoum, Sudan, mimeographed.

Parsons, Kenneth. 1965. "The Tunisian Program for Co-operative Farming". *Land Economics* XLI, November.

_____. 1984. "The Place of Agrarian Reform in Rural Development Policies", in *Studies on Agrarian Reform and Rural Poverty.* Rome: FAO.

Pascon, Paul. 1977. *La Houaz de Marrakesh.* Rabat.

_____. and M. Ennaji. 1986. *Les Paysans sans Terre au Maroc: Connaisance Sociale.* Rabat: Les Éditions Tubkal.

Perkins, Kennel. 1989. *Historical Dictionary of Tunisia.* London: The Scarecrow Press.

Pfeifer, Karen. 1985. *Agrarian Reform Under State Capitalism in Algeria.* Boulder, Colorado: Westview Press.

Raggam, Zouaoui. 1967. "Agrarian Reform in Algeria", in M. Riad El-Ghonemy, ed., *Land Policy in the Near East.* Rome: FAO.

Radwan, Samir. 1969. *Agrarian Reform and Rural Poverty: Egypt, 1952-1975.* An ILO study. Geneva: ILO.

_____, and Anne Thomson. 1981. "Aid Memoire on the Food Subsidy System in Morocco". An IDS and ILO joint study. Geneva: ILO.

_____, and Eddy Lee. 1986. *Agrarian Change in Egypt: An Anatomy of Rural Poverty.* London: Croom Helm.

_____, Vali Jamal and Ajit Ghose. 1991. *Tunisia: Rural Labor and Structural Transformation.* London: Routledge.

Richards, Alan. 1982. *Egypt's Agricultural Development, 1800- 1980: Technical and Social Change.* Boulder, Colorado: Westview Press.

Roemer, John. 1982. *A General Theory of Exploitation and Class.* Cambridge, Mass: Harvard University Press.

Ruttan, Vernon. 1975. "Integrated Rural Development Programs: A Skeptical Perspective", *International Development Review* 17: pp. 9-16.

_____. 1984. "Integrated Rural Development Programs: A Historical Perspective", *World Development* 12: pp. 393- 401.

Saab, Gabriel. 1967. *The Egyptian Agrarian Reform, 1952- 62.* Oxford, U.K: Oxford University Press.

Sayigh, Yusif. 1978. *The Economics of the Arab World Development Since 1945.* London: Croom Helm.

_____. 1991. *Elusive Development: From Dependence to Self-reliance in the Arab Region.* London: Routledge.

Scobie, Grant. 1981. *Goverment Policy and Food Imports: The Case of Wheat in Egypt.* A Research Report No. 29. Washington D.C: IFPRI.

Shaaeldin, Elfatih. 1987. *The Evolution of Agrarian Relations in The Sudan: A Reader.* The Hague: Institute of Social Studies.

_____. and Richard Brown. 1988. "Toward an Understanding of Islamic Banking in Sudan: The Case of Faisal Islamic Bank", in Tony Barnett and Abbas Abdel-karim, eds., *Sudan: Capital, and Transformation.* London: Croom Helm.

Siddiqi, Muhammad. 1981. *Muslem Economic Thinking: A Survey of Contemporary Literature.* Leicester, U.K.: The Islamic Foundation.

Singh, Ajit. 1987. "The World Economic Crisis: Stabilization and Structural Adjustment", *Labor and Society* II.3, September. Geneva: ILO.

Staatz, John. 1990. "Measuring Food Security in Africa: Conceptual, Empirical, and Policy Issues", *American Journal of Agricultural Economics*, December.

Streeten, Paul with Shahid Burki, Mahbubul Haq, Norman Hicks, Frances Stewart. 1981. *First Things First: Meeting Basic Needs in Developing Countries.* Oxford: Oxford University Press.

Svedberg, Peter. 1984. *Food Insecurity in Developing Countries: Causes, Trends, and Policy Options.* Geneva: UNCTAD.

Swearingen, Will. 1988. *Moroccan Mirages: Agrarian Dreams and Deceptions, 1912-1986.* London: Tauris & Co. Ltd.

Tantawy, Mohamad Sayed. 1992. *al-halāl wal-harām fi mo'amalat al-bonouk.* Cairo: al-Ahrām al-Iqtisadi.

Teklu, Tesfaye, Joachim von Braun and Elsayed Zaki. 1991. *Drought and Famine Relationships in Sudan: Policy Implications.* Research Report 88. Washington, D.C: IFPRI.

Tetzlaff, Rainer. 1984. "Kenana: The Biggest and Most Costly Sugar Plant in the World", *Afrika*: 25, No. 2, pp. 11-3.

Tunisia, Ministry of Agriculture, *Annuaire des Statistiques Agricoles*, various issues. *Enquête Agricole de Base*, 1976 and 1980. Tunis: Ministère de l'Agriculture. "Socio-Economic Indicators for Monitoring Agrarian Reform and Rural Development". Institut National de Statistiques, mimeographed in Arabic.

Tully, Dennis, ed. 1990. *Labor, Employment and Agricultural Development in West Asia and North Africa*. An ICARDA study. Dordrecht, the Netherlands: Kluwer Academic Publishers.

Tyler, Godfrey, Riad El-Ghonemy and Yves Couvreur. *Forthcoming*. "Alleviating Rural Poverty Through Agricultural Growth", *The Journal of Development Studies*.

UNDP. 1991 and 1992. *Human Development Report*. New York: Oxford University Press.

United Nations. 1980. *Patterns of Urban and Rural Population Growth*, New York.

_____. *National Account Statistics*, several issues, New York.

United States AID. 1986. *Policy Determination of the Agency for International Development: Land Tenure*. No. PD-13, Washington D.C.

United Nations Fund for Population Activities (UNFPA). 1979. *Survey of Laws on Fertility Control*, New York.

Uzair, Muhammad. 1981. "Some Conceptual and Practical Aspects of Interest-Free Banking", in Ahmad Khurshid, ed.

Van der Hoeven, Ralph and Michael Hopkins. 1981. *Economic and Social Policy Synthesis Program*. Geneva: ILO.

Von Braun, Joachim and Hartwig de Haen. 1983. *The Effects of Food Price and Subsidy Policies on Egyptian Agriculture*. Research Report No. 42. Washington, D.C: International Food Policy Research Institute (IFPRI).

Wally, Youssef. 1982. *Strategy of Egyptian Agricultural Development in the Eighties*. Cairo: Ministry of Agriculture.

Wazzan, Salah. 1974. "Evaluation of Agricultural Land Consolidation of Holdings in Newly Reclaimed Land in Egypt". Alexandria, Egypt: UNDP, mimeographed in Arabic.

Weexsteen, Raoul. 1977. "Révolution Agraire et Pastoralisme", in Bruno Etienne, ed. *Problems Agraires au Maghreb*. Paris: Centre National de la Recherche Scientifique.

World Bank. *World development Report*, several years. Oxford: Oxford University Press.

_____. 1980. *Tunisia: Social Aspects of Development*. Washington, D.C.: The World Bank.

_____. 1986. *Poverty and Hunger: Issues and Options for Food, Security in Developing Countries*. A World Bank Policy Study, Washington, D.C.

_____. 1988. *Rural Development: World Bank Experience, 1965-1986*. A World Bank Operations Evaluation Study. Washington, D.C.

_____. 1991. *Egypt: Alleviating Poverty During Structural Adjustment*. A World Bank Country Study. Washington D.C.: The World Bank.

Wright, Robin. 1992. "Islamic, Democracy and the West", *Foreign Affairs* 71: pp. 131-45.

Zahlan, Antoine and Wadie Magar, eds. 1986. *The Agricultural Sector of Sudan*. London: Ithaca Press.

Index